The Content of Science

The Content of Science:
A constructivist approach to its teaching and learning

Edited by

Peter J. Fensham
Richard F. Gunstone Richard T. White

Routledge
Taylor & Francis Group
LONDON AND NEW YORK

First published 1994 by Routledge Falmer

2 Park Square, Milton Park, Abingdon, Oxon OX14 4RN
711 Third Avenue, New York, NY 10017, USA

Routledge is an imprint of the Taylor & Francis Group, an informa business

First issued in hardback 2016

A catalogue record of this publication is available from the British Library

ISBN 978-0-7507-0220-1 cased
ISBN 978-0-7507-0221-8 (pbk)
ISBN 978-1-138-16580-9 (hbk)

Library of Congress Cataloging-in-Publication Data are available on request

Typeset in 9.5/11 pt Times by
Graphicraft Typesetters Ltd., Hong Kong

Contents

Figures and Tables

Figures

Tables

Preface

Most of the authors in this book are science educators, a professional group that can almost be said to have been created by the wave of curriculum reforms in school science that occurred in the 1950/60s.

The curriculum projects of that period were staffed by academic scientists and by experienced teachers. The former determined what should be taught — the content — and the latter used their experience and some rather global ideas from psychology to present this content in the way it should be taught. As the projects finished, a number of the teachers who worked on them became the next generation of professionals involved in science teacher education. The experiences of being involved in curriculum development, of seeing their ideas tried by other teachers and their curricular packages used by schools and whole school systems, gave them different interests and skills from their predecessors as teacher educators. They had new questions to ask and to answer about science teaching and learning. They now had the time and the responsibility, as well as access to resources, to develop new, well-formulated bases for use in science curriculum. They and their successors also began research on many aspects of the teaching and learning of science. Some of them identified with various social movements, and worked to understand and to translate them into terms that made sense in science curricula.

Throughout the 1980s the research of many of these science educators emphasized ways in which scientific phenomena are experienced and conceived by learners of various ages. Some of the authors in this book have been prominent among those presenting the findings of this work to wider audiences. Rosalind Driver's *The Pupil as Scientist* in 1983 and her book with Edith Guesne and Andrée Tiberghien, *Children's Ideas in Science* in 1985 are examples. In 1988 Peter Fensham's *Development and Dilemmas in Science Education* presented a broad range of views on this research and on the meaning of the social movements and their relation to science education. In the same year, Richard White's *Learning Science* appeared in which the links between these personal conceptions and learning science in formal schooling were explored extensively. Subsequently, White and Gunstone in *Probing Understanding* made widely available the research methods that have been so powerful in elucidating these learner responses to science. Cliff Malcolm, as early as 1987, in *Science Frameworks* drew the research findings into the practical arena of curriculum development at the school and teacher levels. Driver and her Leeds colleagues also at this time performed a similarly practical task with *Approaches to the Teaching of the Particulate Theory of Matter* (CLIS, 1987).

The last of these books is about teaching a particular content area in science. It is that problem that we, in 1991, decided was the next phase of our research that should be popularized. Accordingly, we invited fourteen science educators (ten from centres outside Australia) each to draft a chapter on the implications that the research studies in a specific content area of science have for its teaching. We also challenged teachers who had been meeting regularly for a number of years as the Monash Children's Science Group to participate. Three primary teachers and one secondary biology teacher responded. Kate Brass, an experienced research assistant, was made available to facilitate this uncustomary task for the teachers of writing an account of part of their teaching.

We did not assign particular content to anyone, although in choosing the participants we did consider that it was important at least to have represented content from each of chemistry, physics and biology. We also wanted to include some of the newer views of science content. The eighteen resulting draft chapters were distributed a few weeks before all the authors gathered at Monash for ten days to review them. For once in our academic lives, we had the luxury of time — half-a-day was available for each paper. Two persons had been asked to attend specially to each paper and they usually opened up the discussion of its ideas. After half-an-hour or so, the author was given a chance to respond to the range of initial critiques before the discussion moved to the points others wished to raise. Towards the end of each session, the focus of the discussion moved to suggestions for the revision of the chapter.

On the last day of this Writing Workshop, the threads of this exhilarating period were pulled together as we addressed the issue of the papers becoming chapters in this book. There was a strong feeling we should aim for more than just the set of revised papers. As constructivists, we should attempt to allow the processes of construction we had just been through to be evident. One feature of this was the emergence of a number of issues that recurred from the papers. We have attempted to present some of these issues in the first introductory chapter, not in any sense to resolve them, but rather to share them as constructions of importance from us as a group of teachers and researchers of science education.

We were conscious that constructivist approaches to teaching and learning have long had a place in the theory and practice of pedagogy. Accordingly, we suggested to David Hawkins that he write a short piece that affirms our respect for those on whose shoulders we are now trying to stand for the sake of better science education. His piece forms a second part of the Introduction.

In keeping with our recognition of complexity as an important aspect of science and science education, we found there were multiple sequences in which we might have presented the papers as chapters. Most of the papers moved regularly and easily along a theory-practice dimension, although some are located more consistently towards one or other of these ends. Few of the papers were so concentrated in content terms, or in their approach to the teaching of content, that they could obviously be located together. The sequence we finally chose is thus only one of a number of possible orderings that we discussed in the Workshop.

Peter Fensham (chapter 2) and Merl Wittrock (chapter 3) come at the beginning and provide approaches to curriculum analysis that are pedagogically based rather than related, as has often been the case, to ideas of the structure of the disciplines. Then follow eight chapters in which the authors deal with a variety of forms of science content that go beyond factual and conceptual content. To say these 'go beyond' rather than are 'other than' facts and concepts, is meant deliberately to imply that each of these authors maintains the importance and role of facts and concepts in the particular content they are emphasizing as the knowledge worth learning. Mike Watts (chapter 4) relates problem solving to conceptual learning, while Cliff Malcolm (chapter 5) brings out the importance of conceptual learning in the processes younger learners would undergo in tackling a major problem task involving science and technology. Gaalen Erickson (chapter 6) discusses the manner in which conceptual understanding influences how primary age learners approach the empirical evaluation of various commercial products. Kate Brass, with Maureen Duke (chapter 7), Wendy Jobling (chapter 8) and Telsa Rudd (chapter 9), then provide fascinating accounts of primary teachers and their pupils' explorations of *soil*. This common topic for elementary science illustrates the complexity of science content as each chapter opens up different dimensions of substantive and syntactical science. Each, however, illustrates these teachers' commitment to applying constructivist learning principles. Dick Gunstone (chapter 10) raises the issue of metacognition as essential content of science education, and finally Malcolm Carr *et al.* (chapter 11) illustrate with the topics of *floating and sinking*, *energy*, and *photosynthesis*, how a number of fundamental aspects of the nature of science constitute important content.

Jim Wandersee (chapter 12) uses micrographs in biology (an example of the common problem in science of two dimensional images of three dimensional objects) to point to some principles of pedagogical content knowledge. The account by Kate Brass with Anne Symons and Susan Odgers (chapter 13) of teaching *The Body* follows as a nice example of teachers putting such principles into practice in their high school biology classes.

The next four chapters more directly address the teaching of familiar content topics. Reinders Duit and Peter Haussler (chapter 14) consider *energy* and some of its societal applications. Rosalind Driver and her colleagues (chapter 15) look at the teaching of the phenomenon of *rusting*, Ruth Stavy (chapter 16) discusses research findings about *solids* and *liquids* and what they imply for teaching these *States of Matter*, a topic also addressed in chapter 2 with respect to the *gaseous state*. Laurence Viennot (chapter 17) looks at problems of teaching *Newton's Second and Third Laws*, and at those numerous situations in science where multivariable problems occur.

Richard White (chapter 18), in a paper that seemed appropriate either as the first or the last chapter, discusses the need for a theory of content that is related to the tasks teachers face in making pedagogical choices about content.

References

CLIS (1987). *Approaches to the teachings of the particulate theory of matter*. Leeds: Centre for Studies in Science and Mathematics Education, University of Leeds.

DRIVER, R. (1983). *The pupil as scientist?* Milton Keynes: Open University Press.

DRIVER, R., GUESNE, E. and TIBERGHIEN, A. (Eds.). (1985). *Children's Ideas in Science*. Milton Keynes: Open University Press.

Fensham, P.J. (1988). (Ed.). *Development and Dilemmas in Science Education*. London: Falmer Press.
Malcolm, C. (1987). *The Science Framework P10: Science for Every Student*. Melbourne: Ministry of Education, Victorian Government.
White, R.T. (1988). *Learning Science*. Oxford: Blackwells.
White, R.T. and Gunstone, R.F. (1992). *Probing Understanding*. London: Falmer Press.

Chapter 1 Introduction

Part I. Science content and Constructivist Views of Learning and Teaching

Peter J. Fensham, Richard F. Gunstone and Richard T. White

The first part of this introductory chapter lays out the issues that the group discussed on the final day of the Workshop. These issues fall into two broad groups, those associated with content and those with constructivist views of learning.

Content

When we began planning the Workshop that led to this book, we thought it should be about how the nature of science content affects its learning and teaching. We now see that this was far too simple a notion. What matters is how we think about content, and about learning and teaching, and there is not a single causal path: we knew that our views of learning affect our teaching, but now we see that they also affect our perceptions of content. The chapters that follow illustrate these interactions.

Views of content, learning and teaching not only affect each other, but are driven by changing social forces. The perception in the 1950s that the United States was falling behind the USSR in the space race spurred fundamental changes in American science curricula that soon affected other Western countries. Industrial applications and historical developments of ideas disappeared from the curriculum to make room for more conceptional and theoretical aspects of science (Fensham, 1976). Schools currently are coping with another social change, of having to provide science as a meaningful study for all students, rather than the small minority who might become the next generation of professional scientists. This change makes concentration on theory less appropriate than a focus on the relation between science and technology and social questions such as the balance between the environment and the maintenance of industrial civilization.

At different times different purposes for science education come into favour. The most visible effect of change in purpose is new content but a more subtle and more fundamental effect is a shift in the emphasis placed on the curriculum. Roberts (1982) identified seven curriculum emphases, each of which conveys a different message about the nature of science. One emphasis, for example, is on the processes of science (exemplified by the curriculum, *Science — A Process Approach*), which contrasts with another on the products of science in which students focus on the outcomes of the processes, the description of natural phenomena that scientists present. Different interest groups favour different emphases. For example, leaders of the Science-Technology-Society (STS) movement (for example, Aikenhead, 1986; Malcolm, 1987) favour three of Roberts' emphases to do with coping with one's environment, applying science and technology to social problems, and providing a solid foundation for further learning. Advocates of the need for school science to become more sensitive to the interests of girls (for example, Kahle, 1985; Kelly, 1987; and the McClintock Collective, 1987), emphasize science content that includes several of Aikenhead's categories of STS learning (*Women in science and technology, Scientists and their personal traits, Motivation of scientists, Social nature of scientific knowledge*, and *Social responsibility of science*), but adds more personal and subjective responses to nature and to society as important dimensions.

Hence emphases, and thus the content, change as one group or another becomes influential. Some interests may hold sway at one level of schooling while others may wield the influence at another level. Millar and Driver (1987) discussed the particular polarization of this sort that emerged strongly in a number of school systems following the 1960/70s reforms. Secondary schools concentrated on product, primary schools on process.

These reconceptualizations of the emphases and content of school science have not merely been academic debate and discussion. Almost all curricular developments in the last decade have responded in varying degrees to these new ways of defining content and refer often to 'concepts in contexts'. Different curricula have seen the nature of these 'contexts' in different ways, thus resulting in a variety of curricular emphases. For example, Salter's Chemistry and Salter's Science in Britain include many *applications of science* as the contextual starting point for the content to be learnt (Smith, 1988). The PLON material for physics in the Netherlands emphasises *Science as a way of knowing* and *STS decisions* (Eijkelhof and Kortland, 1988). A Canadian secondary course for grades 9 or 10, *Logical Reasoning in Science and Technology* also sets socio-scientific contexts and the decisions they invoke as very central in its range of science content (Aikenhead, 1991). The McClintock Collective (1987) in Australia has produced curriculum materials that have as content the gender inclusive aspects of science referred to above.

An important outcome of this lively debate about the content for school science has been the recognition that polarized solutions like the one mentioned earlier are no solution at all. While consideration of topics leads to different conclusions from those derived from consideration of emphases, these and other aspects of science are not discrete. They interact with each other. The measuring instruments scientists use embody conceptual relations established in earlier stages of scientific enquiry. Observations are not independent of criteria that stem from categories for conceptualizing the phenomenon being observed. Likewise, school science education is not helped by trying to isolate categories of content. When the connections between the categories are ignored, the distinctions between them tend to be overlooked. For example, if how concepts are invented in science is not part of the content to be learnt, it is not surprising that teachers and their students see *Chlorine is a green gas* and *Chlorine has a covalent bond* simply as two equivalent pieces of chemical knowledge worth one mark each if a student remembers either or both. The profound epistemological distinction between them is lost on both teachers and students.

These relations between social forces and perceptions of purpose and content lay behind discussions in the Workshop, and influenced the emergence of three major issues concerning science content: its variety, its complexity, and the relation between content and action.

Variety of Science Content

Roberts' emphases, Aikenhead's categories and other dissections of purpose and content demonstrate the richness of science as a human activity, and hence of its potential to provide meaningful content for all students to learn at school and, indeed, for the enrichment of the lives of all citizens. A classification of curriculum content like the commonly used *knowledge, skills and attitudes* is both too simple and too abstract to do justice to this richness in science. The simplicity of this classification pushes into its three categories aspects of the human exploration of nature that are epistemologically different. Its abstractness divorces science content from the dynamic and human situations of its origins and its learning. It is now time to replace, in curriculum thinking and planning for science, this unhelpful and oversimplified trilogy for the content worth learning. New typologies for describing science content are needed, and those that are emerging need to be encouraged in the debates about science for the school curriculum. Because of the complexity of science (see below), no typology will be ideal or even pragmatically the neatest. Rather, some will be more helpful and useful than others, depending on the context or the stage of debate that the teachers and curriculum decision makers are in.

Accordingly, it is useful to look at Roberts' emphases as a seven-fold typology in relation to the science content represented in the eighteen chapters of this book. A number of the authors address several different sorts of content and relate to more than one of Roberts' emphases. Of the many examples of explicit content used by the authors, most seem, without much distortion, to be associated with one or other of the seven emphases, and each emphasis has at least two authors addressing its content. White (chapter 18) is an exception. This is not surprising since he is also addressing the typology question when he addresses the need for a theory of content, but from a pedagogical perspective. On the other hand, if the *Pedagogy of Science* were to be added to Roberts' list of emphases then the content of White's chapter would be an example of it as would also be some of Gunstone's (chapter 10) and Carr's (chapter 11) content. There is a strong argument that *Pedagogy of Science* is an essential ingredient of science and even more so for it to be a category for content in science education. Jenkins and his colleagues in Alberta were so convinced about this that they added *Communication* as a fourth dimension to the usual S-T-S when they published their curriculum materials for senior chemistry (Jenkins *et al.*, 1989).

Complexity of Science Content

The interdependence within science content is demonstrated in the examples discussed by the authors. The four chapters (7, 8, 9 and 13) by the school teachers perhaps best illustrate this interdependence. The primary teachers, for example, easily transcend the powerful boundaries in professional science and in secondary school science that are called the science disciplines. One of the teachers starts with astronomical issues but is soon into micro-structures of soil. Another slides from soil textures to the dynamic organic/inorganic composition of soil, and on to socio-technical decisions. The teachers addressed whatever disciplinary content was relevant to their attempt to be faithful to the phenomenon under study and to their attempts to investigate it with their students.

The contents of science addressed by the researcher authors were less fluid than those of the teachers in these interdisciplinary and practical senses, but they do point to other complexities of the content of science. One that several authors discuss is the pair of levels in science (and hence in science education) at which phenomena are explored and described. There is the macroscopic level at which materials and organisms are physically observed and handled, and there is the microscopic level or atomic/molecular level at which explanatory descriptions and powerful ideas can be provided. In the practice of the sciences, thanks to the invention of more and more powerful extensions of the human senses, empirical investigation at both these levels goes on. Teachers and learners in schools are rarely able to engage in investigation at the micro or atomic level and so their experience of this level of scientific discourse differs from what they can achieve at the macro level (see chapters 12, 13 and 16). Unfortunately, as several of the authors point out, learners' experiences of phenomena at the macro level rarely seem to need discontinuous or kinetic molecular models of matter. Furthermore, the properties of the building blocks at the atomic level, as usually presented in school science, are too over simplified or idealized to account for the gross, collective properties that are observed at the macro level. Here is a quite fundamental complexity in science that science education is only just beginning to recognize. It is certainly now well established as a very confused aspect of the teaching and learning of science.

Science Content and Action

Successful students in school science accumulate a great deal of passive knowledge, but often surprisingly little of what Layton (1992) calls knowledge for practical action. So perhaps science content in relation to practical action is what really needs to be considered if the phrase 'concepts-in-contexts' (see above) is to become clear to teachers and their students as a call for a quite new and distinctive type of learning outcome in science. The contexts for action are also important for the form that knowledge takes and hence how it appears as content in science education. The student who has mastered the concept of volume in terms of cubic decimetres, the official scientific unit, will flounder in the every day world where most measures of volume are in terms of litres, or cups, or wheelbarrows.

A number of the authors in this book raise examples of science knowledge in relation to a variety of human actions. These human actions may be attempts to teach or to learn the knowledge, applications of the knowledge in practical or technological situations, or explorations at a deeper level of the knowledge itself. The relation between what is known and the knower can involve both facilitation and constraint. What primary students know about magnets, as Erickson (chapter 6) shows, determines to a considerable extent how they will go about a task involving magnets. An action task on the other hand, as Watts (chapter 4) shows, can establish contexts in which conceptual knowledge is eagerly sought and learnt in some depth. The school teachers' chapters are again particularly good examples of the reflexive and iterative nature of relationships between knowledge and actions. One feeds on the other. To realize the potential this relation has for learning requires the learners to be aware of it, a point taken up by Wittrock (chapter 3) and Gunstone (chapter 10). When they are observing or measuring with an instrument in a science experiment (as Carr *et al.* in chapter 11 discuss), they need to know they are using as points of reference concepts and relations that have been so thoroughly established in earlier phases of science that we now trust them as tools.

Constructivist Teaching and Learning

At the same time as the Workshop participants were discussing aspects of content, we were talking about teaching and learning. Although we were aware of the differences between types of content and emphases that could alter between curricula, and we referred to different procedures of teaching, we tended to think in terms of only one model of learning. Perhaps we should have thought more about whether that was appropriate.

Discussion of the papers in the Workshop rapidly made clear that all participants subscribe to a constructivist view of learning, with its fundamental principle that people construct their own meanings for experiences and for anything told them. The constructed meaning depends on the person's existing knowledge, and since it is inevitable that people have had different experiences and have heard or read different things, all have different (though often similar) meanings for any concept. Since this principle is universal, the Workshop participants must, accordingly, differ in the meanings they have for construction itself, and so with readers of this volume. We have our own understandings of what is involved in learning science.

Difference does not mean opposed or unrelated. The papers and the discussions in the Workshop revealed much shared meaning. That was not inevitable; for even though invitations to the Workshop might be expected to go to a congenial group, the participants came from eight countries that are far apart geographically and could well have developed conflicting notions of construction. There were differences, but these were not so great as to impede discussion. Instead, discussion probably brought participants' meanings to be more similar, as well as, one would hope, richer and more subtle.

One point to be considered was the use of the terms 'constructivist learning' and 'constructivist teaching'. Since all learning involves construction, we argued, 'constructivist learning' is tautologous. Some participants also wanted to avoid 'constructivist teaching', preferring to refer to teaching that takes account of the principle that each learner constructs personal meaning from experience and instruction. Others found the shorthand term convenient and, while agreeing with much of the logic of those arguing against the term, believed it was helpful in drawing teachers' attention to the need to consider changes in their pedagogy.

Not surprisingly, this difference remained unresolved as the issue was not the semantic appropriateness of the term but rather how useful 'constructivist teaching' was seen to be as a label for initial focussing of teacher thinking.

In discussion, the participants were clear that construction does not give students licence to claim that their meaning is as good as scientists' meaning, no matter what its form. Construction does not mean 'anything goes'; some meanings are better than others. Means for determining what is better are then significant. Carr *et al.* (chapter 11) propose that the criteria for an explanation of a natural phenomenon are that it should be elegant and parsimonious and connected with other phenomena, as well as having the intelligibility, plausibility and fruitfulness suggested by Posner *et al.* (1982). We now add that the explanation should be testable. For many people creationism provides a powerful explanation of the world, which no doubt they find elegant, parsimonious, connected, intelligible, plausible and fruitful; but it relies on faith and is not testable, so is not science.

Participants also distinguished construction from discovery. All learning involves construction of meaning, whether the knowledge is discovered or received by direct transmission. We recognized a distinction between the pedagogical role of the teacher in construction and discovery and hence between the teaching we are advocating and what has often been advocated for discovery learning in science education. Construction can be guided, to a greater or lesser or more subtle extent. Often, for efficiency or so the learners' meanings will be more likely to meet the criteria of Carr *et al.*, there will be much guidance. Teaching involves judgments about how much and what form of guidance is best for any topic and any group of learners, and when to provide it. Viennot suggested a metaphor which was much used in the Workshop, of the teacher 'parachuting in'. This useful image distinguishes parachuting from 'free fall' (the teacher landing heavily on the students' views, squashing them underground) and from 'floating endlessly above the surface' (leaving the students to make whatever sense they like). To parachute is to drop lightly but effectively on the appropriate place at the appropriate time. Judging when and where to do it is an advanced skill, requiring the teacher to have both pedagogical and content knowledge. As the teachers in chapters 7, 8, 9 and 13 and Scott *et al.* in chapter 15 describe the practice of these skills, it is pedagogy that is learner centred but teacher controlled in a way that there is always something the learners are called on to construct. This is the most significant of the differences between the discovery learning approaches advocated in the 1960s and the constructivist-based approaches considered in this volume.

Conceptual Replacement or Addition

The purpose of intervention by the teacher is to bring students' conceptions to fit criteria such as those of Carr *et al.* — that is, to promote conceptional development and change. Much writing about conceptional change is in terms of abandonment of primitive views and acceptance of a sophisticated, scientific one. A theme that emerged in the Workshop, however, was that conceptional change is only rarely a sharp exchange of one set of meanings for another, and is more often an accretion of information and instances that the learner uses to sort out contexts in which it is profitable to use one form of explanation or another. Thus scientists will speak, and think on occasions, of sucking liquid up in a straw rather than of creating differences in air pressure; will not be concerned by signs such as 'No animals allowed on freeway' even though they know they are members of the animal kingdom; and accept that in the day-to-day world things stop moving if no force is applied to them, whatever Galileo and Newton said. They have learned which meaning and which principle fit each context. Such addition of conceptual meaning is gradual, and does not all occur in formal lessons. Concepts are revisited, and the meanings build and shift a little each time.

The example of drinking through a straw illustrates the point about shifts as well as additions in meaning. When learners come to understand the notion of pressure difference, they do not drop the word 'suck', though their conceptions of sucking change. Knowledge about pressure has been added, but old knowledge is revised rather than abandoned. A conceptual addition has occurred. Central to this formulation of what is often described as 'conceptual change' is that the individual also has informed approaches to deciding which of a number of meanings is appropriate in a particular context.

Addition and revision of meaning involves perceiving links between concepts. The Workshop participants frequently observed that concepts vary in how linkable they are, though all are linkable to some extent. Good learning incorporates linking, and good teaching promotes it. Even better learning follows when students comprehend why links are important, and actively seek them for themselves between topics and across subjects. Such behaviour is a characteristic of metacognition, which the Workshop participants recognized as consistent with constructivism. They referred frequently to the central role in conceptional change of learners' awareness and control of their thinking. But construction is not confined to student learners. Teachers, too, construct their own meanings. In relation to their content and pedagogical responsibilities, it follows that teachers need also to be aware of themselves as learners, and that their constructions of meaning for these are never complete.

Situated Cognition

In addition to metacognition, the Workshop discussions touched on the relations between conceptual change and the notion of situated cognition, and on the phenomenographic interpretations of Ference Marton and his associates at the University of Gothenburg.

Situated cognition (Brown *et al.*, 1989; Lave, 1988) points out that all cognition is situated in the context of the activity associated with the learning. 'The activity in which knowledge is developed and deployed . . . is not separable from or ancillary to learning and cognition. Nor is it neutral. Rather it is an integral part of what is learned. Situations might be said to co-produce knowledge through activity' (Brown *et al.*, 1989, p. 32). This argument, that what is learned is inseparable from how it is learned and used, has links with the position advanced in this workshop that understanding the contexts of application of different personal conceptions is central to understanding of science.

The Workshop did not, however, go on to debate the alternatives to conventional teaching that proponents of situated cognition advocate. Nor did we have time to explore the consistency of conceptional change as accretion, restructuring and linking with the Gothenburg group's notion of related conceptions, which holds that not only are multiple conceptions held by an individual and differentially used in different contexts, but that these multiple conceptions are related and can be shown to have a form of hierarchy of inclusiveness. Although the views that emerged at the Workshop may turn out to differ in important respects from situated cognition and the Gothenburg notion, there was sufficient convergence to encourage further discussion of their implications for classroom practice.

Despite the subscription and enthusiasm of the Workshop participants for constructivism, we were conscious that we had not invented it. Neither is it merely the fashionable band wagon of the moment on to which we have clambered. To acknowledge the heritage of the ideas we believe are so important for teaching the content of science, we asked David Hawkins, our senior member, in Part II of this Introduction to provide a brief history of this heritage.

Part II. Constructivism: Some History

David Hawkins

The aim of science is to extend our experience and reduce it to order.
(Niels Bohr)

The learning of science by children and older students, no less than scientific research itself, is in its own way an investigative, constructive process. Some modes of teaching can further this process, others inhibit it. Some reconstruction of subject matter for optimal access by diverse students may also prove essential. The papers in the present volume cluster around these common theses. They seek, in various contexts, to define conditions that promote optimal student inquiry, and thus also the teaching arts that can provide those conditions. The general philosophy that supports this view has come to be called constructivism. The name is new, the substance is not. But the substance needs to be further 'extended and reduced to order'.

Constructivist views do stand in opposition to long-established majority traditions in education generally, especially education in the sciences. In that tradition the dominant image has been the transmission, from teacher and text to students, of single-track, logically organized knowledge. Sometimes, therefore, 'Constructivism' may tend to be taken up less as practice and theory still under development, than as an exclusionist manifesto.

As theory, constructivism has had two major historical sources. One source is philosophical, a general theory of knowledge that can provide background and support for more specific educational theory and practice. The other source is the experience of reflective practitioners, teachers and those who seek to help and learn from them. A third source, growing in recent times, is a professional research community, seeking to bring theory and practice more coherently together. That interest animates the present volume.

Constructivist theory, as we understand it here, is as old as our traditions. Plato learned it by following the teaching practice of Socrates. Socrates taught, for the most part, by insightful questioning that helped others 'reduce to order' their own still-fragmentary knowledge. In the *Meno*, Socrates induces an untutored slave boy, by honest questioning, to establish the Pythagorean Theorem.

For modern times I choose Immanuel Kant as the first major precursor. Scientific knowledge, he says, is actively constructed from our observational experience. For Kant the metaphor of construction is pointedly appropriate. He characterized rational inquiry as necessarily following a certain architectonic that shapes the process. His organization of categories — a basic system of questions that inquiry must ask of nature — guides us in an ongoing process of constructing, testing and reconstructing explanatory hypotheses. The essence of this constructive process, as Kant perceived it, is not easy to distil today. His eighteenth-century argument, abstract in style and rarely illustrated, is difficult even by classical German standards. He was thoroughly acquainted with the history of science, and I can remember one telling historical reference he made to the researches of Stahl and of Torricelli (Kant, 1983). I shall suggest instead some quite Kantian delineations of the investigative art to be found in the writings of three American philosophers.

The first of these is the logician, Charles Saunders Peirce (1839–1914), who wrestled endlessly with Kant's first Critique. In emphasizing the constructive character of scientific knowledge, Peirce found a gap in the traditional account of *deductive* and *inductive* reasoning, and developed the theory of what he called *abduction*, the art of moving from novel phenomena to hypotheses that would, if confirmed, explain the phenomena; then — most importantly — seeking independent evidence that would confirm, or require us to modify or reject them. Peirce was the founder of Pragmatism, a philosophy organized around a somewhat more general principle he enunciated, that we elucidate our beliefs by defining their predictable consequences. Peirce is seldom referred to in educational research literature. There is a good single-volume selection of his vast writings (Buchler, 1955). A careful recent treatment is that of Christopher Hookway (1985).

One of Peirce's associates was William James (1842–1910), whose two-volume *Principles of Psychology* should be considered a major precursor of what some now call cognitive psychology (James, 1901). A second and slightly younger psychologist and philosopher was John Dewey (1859–1952), known in educational circles mainly for his association with the progressive education movement, of which I shall speak below. Because of widespread conventional disparagement of that movement, early and late, Dewey's work has suffered from contumely and neglect. His much later major work, relevant to the methodology of science and to constructivist thought, is his *Logic* (Dewey, 1938). I shall return to Dewey below, as one of the few early theorists of education whose thinking was deeply influenced by his acquaintance with exemplary practice, and who in turn helped guide it.

The acceptance of Kant's view of knowledge as a constructive process led to further issues. If, as he argued, some architectonic guides rational inquiry, then the structure of science does not emerge as that of nature alone; it must also reflect the architectural design. Thus, if — contrary to Kant's own view — different individuals or cultures have different schemas of inquiry, different presuppositions asking nature different kinds of questions, then we will have different systems of knowledge developing! All equally, it would seem, can 'extend experience and reduce it to order'. Far older philosophies of relativism and subjectivism thus found new, 'constructivist' justification. One must mention here Thomas Kuhn's (1970) well-known work in which he carefully analyzed historical shifts in the presuppositions of scientific investigation and thought.

A contemporary explicit proponent of 'constructivist' relativism, known to educational researchers, is Ernst von Glasersfeld (1984). Kant does, indeed, say we cannot know nature as 'the thing in itself', but only as humanly qualified. But he did not intend a great mystery condemning us to our own parochial views; he was simply pointing out — against the rationalistic philosophy of his times — that this prohibition was one of simple logic. It is similarly impossible to determine the colours of logarithms or the pH of justice. Hegel, a follower and critic of Kant, accepted the multiplicity of rational schemes for knowledge, and treated their conflicts as historical drama, motive and as a source of further learning and synthesis.

Another contemporary, known to educational researchers, is Stephen Toulmin (1972). He regards the differences among interpretative schemes in science as improperly taken to justify relativism or subjectivism. Different approaches do indeed develop along the way. This is not a matter for complacency, but for further conceptual evolution — evolution in a more contemporary sense than Hegel's. The history of science provides a good many examples. Such 'alternative conceptions' appear, as we all know, in the thinking of science students. Some may — as in history — generate fruitful disagreements along the way.

It remains to speak of certain minority traditions in education, 'constructivist' in practice and spirit. Some of these go back for at least a century and a half. Mainly confined to education for the years of childhood, they have rarely influenced secondary schooling. There is no comprehensive history, and I shall speak only of some samples I am acquainted with. Little of it is explicitly linked to science education, but much to the enrichment and diversification of classroom environments. The traditions I can speak of developed in England and the United States, inspired early by such innovators as Friederich Froebel, whose own philosophy was Hegelian, essentially constructivist. His guiding principle was the recognition of children's capacity for self-directed learning, *selbstätigkeit*. His 'system' for encouraging it, however, was rigid. Other innovators along the way gradually eliminated these rigidities and developed further the participatory role of teachers. Over time imaginative teachers evolved a large repertoire of classroom material, suitable for a wide variety of engagements and investigations, indoor and out, including much early mathematics and science (see, for example, Cremin, 1961). In the United States this tradition has virtually died out of the public schools since World War II. In England through the 1960s it evolved to major proportions in the infant schools (ages 5–7+), substantially affecting the junior schools (7+–11+), but rarely the grammar schools or beyond. The classic account is the Plowden Report (1967).

At the turn of the century, this movement encountered a major theorist and supporter in the person of John Dewey. I single him out because I believe his best thinking, profoundly constructivist, was guided in part by first-hand acquaintance with experienced and reflective teachers. The school itself, he often argued, was the necessary laboratory in which practice could be distilled as theory, and theory reduced to altered practice. Marching under the banner of Dewey's philosophy, however, 'progressive' schools sometimes earned their disrepute because they fell into the laissez-faire belief that children would pursue their own learning without responsive adult participation. In his last (and I think best) major writing on education, Dewey (1963) recognized that he had been lax in not insistently emphasizing the essential role of investigative teachers: in scouting out the diverse talents of individual children; in recognizing their available pathways of entry into important subject matter; in evolving relevant resources; in furthering children's potential contribution to the vitality of classroom life.

Having spoken of this long and honourable minority tradition, I would like to comment on the important inclusion in the present volume of research reports of work by classroom teachers committed to extend that tradition (see chapters 7, 8, 9 and 13 in this volume). The elementary and secondary school projects of the 1960s and after, of which I was a part, too often reflected the belief that the delineation of rich subject matter and investigative style could be spread across the land by innovators who, like myself, lacked experience as teachers of children or adolescents. We learned much, as four of us have recently tried to explain (Duckworth *et al.*, 1990); but good teachers, more fully involved, could have taught us far more, and added more usefully to their own repertoires. It should be a basic principle of constructivist inquiry to start where the action is.

References

AIKENHEAD, G. (1986). The content of STS education. *A Missive to the Science-Technology-Society Research Network. 2*(3), pp. 18–23.

AIKENHEAD, G. (1991). *Logical Reasoning in Science and Technology*, Toronto: Wiley.

BROWN, J.S., COLLINS, A. and DUGUID, P. (1989). Situated cognition and the culture of learning. *Educational Researcher. 18*(1), pp. 32–42.

BUCHLER, J. (1955). *Philosophical Writings of Peirce*. Selected and edited by Justice Buchler, New York: Dover.

CREMIN, L.A. (1961). *The Transformation of the School: Progressivism in American Education, 1876–1957*. New York: Knopf.

DEWEY, J. (1938). *Logic: The Theory of Inquiry*. New York: Holt.

DEWEY, J. (1963). *Experience and Education*. New York: Collier, Macmillan.

DUCKWORTH, E., EASLEY, J., HAWKINS, D. and HENRIQUES, A. (1990). *Science Education: A Minds-on Approach for the Elementary Years*. Hillsdale NJ: Lawrence Erlbaum.

EIJKELHOF, H. and KORTLAND, K. (1988). 'Broadening the aims of physics education'. In FENSHAM, P.J. (Ed.). *Developments and Dilemmas in Science Education*. London: Falmer Press, pp. 282–305.

FENSHAM, P.J. (1976). Social content in chemistry courses. *Chemistry in Britain*, *12*(5), pp. 148–51.

HOOKWAY, C. (1985). *Peirce*. London: Routledge and Kegan Paul.

JAMES, W. (1901). *The Principles of Psychology*. London: Macmillan.

JENKINS, F. *et al.* (1989). *Science, Technology, Society, Communication — CHEMISTRY*. Edmonton, Canada: Author Group, Karitann Publications.

KAHLE, J.B. (Ed.). (1985). *Women in Science: A Report From the Field*. London: Falmer Press.

KANT, I. (1983). *Critique of Pure Reason*. Preface to the 3rd edition. Translated by Thomas Kingsmith Abbott. London: Longmans, Green.

KELLY, A. (Ed.). (1987). *Science for Girls*. Milton Keynes: Open University Press.

KUHN, T. (1970). *The Structure of Scientific Revolutions*. Chicago, IL: University of Chicago Press.

LAVE, J. (1988). *Cognition in Practice*. Cambridge: Cambridge University Press.

LAYTON, D. (1992). Science and technology teacher training and the quest for quality. In LAYTON, D. (Ed.). *Innovations in Science and Technology Education*. Vol. IV. Paris: UNESCO.

MCCLINTOCK COLLECTIVE. (1987). *The Fascinating Sky*. Melbourne, Victoria: the Ministry of Education.

MCCLINTOCK COLLECTIVE. (1988). *Getting into Gear*. Canberra: Curriculum Development Centre.

MALCOLM, C. (1987). *The Science Framework P-10*, Melbourne, Victoria: The Ministry of Education.

MILLER, R. and DRIVER, R. (1987). Beyond Processes. *Studies in Science Education*, *14*, pp. 33–62.

PLOWDEN, B. *et al.* (1967). *Children and Their Primary Schools: A report of the Central Advisory Council for Education (England)*. London: Her Majesty's Stationery Office.

POSNER, G., STRIKE, K., HEWSON, P. and GERTZOG, W. (1982). Accommodation of a scientific conception: Towards a theory of conceptual change. *Science Education*. *66*(2), pp. 211–27.

ROBERTS, D.A. (1982). Developing the concept of 'Curriculum Emphasis' in science education. *Science Education*. *66*(2), pp. 243–60.

SMITH, N. (1988). In support of an application-first chemistry course: Some reflections on the Salters GCSE scheme. *School Science Review*. *70*(250), pp. 108–114.

TOULMIN, S. (1972). *Human Understanding*. Oxford: Clarendon Press.

VON GLASERSFELD, E. (1984). Introduction to Radical Constructivism. In WATZLAWICK, P. (Ed.). *The Invented Reality: How Do We Know What We Believe We Know?: Contributions to Constructivism*. New York: Norton.

Chapter 2

Beginning to Teach Chemistry

Peter J. Fensham

> Lavoisier in the introduction to his 1793 elementary chemistry text was aware of the problem of constructing meaning in chemistry. 'However certain the facts of any science may be, and however apt the ideas we may have formed about these facts, we can only communicate false and imperfect impressions of these ideas while we want words by which they may be properly expressed'. (Bouma, Sutton and Brandt, 1988)

The first edition of the *Enclopaedia Brittanica* (published between 1768 and 1771 by the Society of Gentlemen in Scotland) devoted 113 pages to '*all the principles of chemistry*'. Chemistry was defined as follows:

> The object and chief end of chemistry is to separate the different substances that enter into the composition of bodies; to examine each of them apart; to discover their properties and relations; to decompose those very substances, if possible; to compare them together, and combine them with others; to reunite them again into one body, so as to reproduce the original compound with all its properties; or even to produce new compounds that never existed among the works of nature, from mixtures of other substances differently combined.

Most chemists today would not be too uncomfortable with this description of their subject, although the number of chemical substances that are now known is many orders of magnitude larger than was the case in 1770. Despite the all pervasiveness of chemical substances, and indeed of chemical changes in the lives of all human societies, chemistry has not proved to be a simple subject to teach in schools.

Although some aspects of chemical phenomena and the properties of some common chemicals may be introduced in the early years of schooling, it is usual to begin to teach chemistry as a discrete field of science at some stage of secondary schooling. In some countries the subject is slowly built up over a number of years, and in others it is more intensively studied in the later years of secondary schooling. Whichever of these patterns pertain, the design of the curriculum for school chemistry (and hence of textbooks to support its teaching) has been problematic.

Satchell (1982), a chemistry professor, addressed the issue of how to begin the teaching of chemistry by arguing against the start that one of his children was receiving in which chemical and physical processes are distinguished, and general names like element, compound, mixture, atoms, molecules, etc. are defined. Instead he suggested starting with a description of atoms and their structure, and proceeding from this to 'intelligible and logical accounts of ions, molecules, elements, compounds, formulae, the laws of constant proportions, etc.'. Lagowski (1985), on the other hand, expressed concern that beginning students are too often presented with principles without any experience of the observations on which they have been based. Gillespie (1976) in Canada and Bucat and Cole (1988) in Australia, academic chemists with abiding interests in school chemistry, argued from the same concern for beginning with chemical reactions. The latter pair had both substantially contributed to a new text, *Elements of Chemistry: Earth, Air, Fire and Water* which, from the first pages, has a focus on chemical reactions rather than on theories of a physical nature such as atomic structures. Two hundred years earlier, the Scottish gentlemen had also concluded that it was more productive to study the reactivity of these elements than '*to tire our minds with vain conjectures about the parts or elements of which they may consist*'. The academic debate just presented has been a periodic one, and texts and curricula that make use of each of these three beginnings for school chemistry are still in use in many countries. Examples are given in the next section.

Each of the three beginnings advocated above is drawn from the structure of the discipline of chemistry itself. Each turns out to have its share of learning pitfalls, and teachers of them often unwittingly confirm Bent's (1981) dictum that '*there is nothing in chemistry that cannot be made harder*'. In my conclusion on page 26, I contrast these approaches with the newer STS chemistry curricula that are emerging. The latter are drawn from the interactions members of society have with substances and their reactions in real life.

Following an elaboration of the three approaches to chemistry, some relevant findings from research on student conceptions will be presented. They then become the basis to discuss research into new approaches to the pedagogy of chemistry from which I draw conclusions about how introductory chemistry curricula should now be designed.

Exemplary Beginnings

The three approaches to the introduction of chemistry can be designated in a shorthand manner as A. Substances; B. Atomic Structure; and C. Chemical Reactions. A and C have been more common when chemistry is begun earlier in schooling, and B has been more common when chemistry is begun later although Satchell was arguing for it for 13–14-year-olds.

A. Substances approach

The introductory module of *Interdisciplinary Approach to Chemistry (I.A.C.)*, a 1970s course in the USA for students of 15 or 16, is an example of the substances approach. It introduced substances as either *mixtures* or *pure*, and distinguishes between *chemical* and *physical changes. Elements* and *compounds* followed, and hence students came to formulae, equations and nomenclature. A contrived cycle of reactions involving copper aimed at a chemical principle of conservation. Solutions, states of matter, and energy and kinetic theory were introduced before chemical reactions became the topic for teaching and learning. Even then, reactions were examples of types of reactions — *acid/base*, *oxidation/reduction*, and *precipitation* — rather than the means of making a new substance with different properties that are desired by chemists or needed by society. The module closed with atomic structure, bonding and molecular shapes.

B. Atomic structure approach

The atomic structure approach was used to develop the materials of CHEM Study, a project of the National Science Foundation for grades 10 or 11 students in the USA. These were very widely used in the 1960s and 1970s as a curriculum in a number of countries, either directly or in adapted forms that often also spread the learning over several of the later secondary years. The architects of CHEM Study essentially began with atomic theory and moved directly to symbolic representations of chemicals and to calculations involving the mole as the unit in chemistry. The abstractions and principles of kinetic theory, phase equilibria, atomic structure and periodicity, chemical equilibrium and reaction rates are all covered before the more descriptive chemistry of the elements and of chemical reactions to produce new substances.

This approach with its strong base in atomic scale conceptions of matter has, in the 1980s, been adding applications of its concepts but now faces serious challenges from the more sophisticated forms that curriculum design under the S.T.S. movement is now taking (see page 24 and Fensham, 1988).

C. Chemical reactions approach

Nuffield 'O' level Chemistry, developed in the early 1960s in England for able students aged 13 to 16, began with the problems of separating pure substances from the mixtures and compounds in which they occur naturally. It used chemical reactions and physical separation processes to solve them. This course then moved straight into reactions in which substances are decomposed by heating in air or burnt with air as the source of the reacting oxygen.

States of Matter: A Central Topic in Chemistry?

In each of these approaches it is usual to find the topic 'Properties and States of Matter' quite early in the sequence of teaching. This topic is interesting because it is one of the two commonly found in introductory chemistry (atomic structure is the other) with which both chemists and physicists identify.

Properties of solids, such as melting point and electrical conductivity, interest chemists because they reflect the nature of the bonding in this state. They interest physicists as phenomena which relate to other properties and uses of these materials. The behaviour of gas under changes of temperature and pressure are of joint interest. In the case of the liquid state, more distinctive interests are again evident. Physicists relate directly to properties of this state such as flow, viscosity, compressibility, etc., whereas the interests of chemists lie with the components that are present in the liquid state. Thus chemistry does consider the liquid state of ionic substances because of the potential in some molten salts for electrolysis. Chemists' major interest in the liquid state is, however, in solutions and in the various degrees of mixing that occur between a solute and a liquid solvent — a topic of no interest to physicists.

The interplay between the states of matter — melting and solidifying, evaporating, boiling and condensing and subliming — seem to follow logically from the two basic assertions in this topic. *Matter exists in three states — solids, liquids and gases. The existence of these states depends on the prevailing conditions of temperature and pressure.* The truth of these assertions for the 104 or so elements is testified by values for the melting point and boiling point and for these conditions in any data book. Scrutiny of these values indicate, however, that for only a handful of these elements can a change of state be observed in school laboratories, or in the changing conditions that occur for life on earth.

These assertions, however, do apply to water, that altogether ubiquitous and familiar substance, the properties of which so uniquely relate to life on earth. It is the regular exemplar that teachers and textbooks use to illustrate them. Among the myriad of other compound substances and mixtures in the elementary data books, it is hard, however, to find other examples for which these assertions hold in their totality. Many of the common substances in introductory chemistry do not undergo one or more of these transitions of state. The decomposition on heating of most nitrates, carbonates and hydrated salts confuses the notion that raising the temperature and controlling the pressure will lead simply to the phenomena of melting, or sublimation and boiling.

Chemistry teachers do regularly deal, in introducing their subject, with pure substances that exist in one or other of these three states. Air, oxygen, nitrogen, and chlorine are gases as are the oxides of carbon, nitrogen and sulphur, and the compounds of hydrogen with carbon, sulphur and the halogens. Benzene, ethanol, bromine, and mercury are liquids, and a host of the inorganic and organic compounds introduced early are solids. Rarely, however, do chemists (or their students) have to consider any of these as pure substances in more than one state. Only in mixed component situations such as separation by distillation or crystallization, and the use of melting points to determine the purity of organic solid substances does chemistry involve two states of matter in an operational sense that students can share early in their studies. At a more advanced stage, chemists and their students are interested in phenomena like the changes of state in solutions, because of the possibility of relating the depression or elevation of the pure solvent's melting or boiling point values to the relative molecular mass of the unknown solute substance.

If states of matter and the transitions between them are so rarely encountered in chemistry, why has the topic been seen as so important to its early teaching? The answer seems to lie in this topic's role in introducing students to the kinetic molecular theory and to the views of what matter is like on the atomic or molecular scale. Moreover, the gaseous state, and transitions to and from it, is the aspect of the topic that is most used to develop these kinetic molecular ideas.

A Comment

I do not wish to suggest that these examples of the approaches to introductory chemistry courses were ignorant of or uninfluenced by educational research. Their main formative influences from research were, however, drawn from epistemological ideas about science and from psychological studies of learning that involved content and processes quite different from chemistry. There was a general belief among these curriculum developers that cognitive learning would follow in a heuristic manner from carefully designed laboratory exercises. Unifying principles were seen to have a powerful and helpful role in learning the profusion of knowledge chemistry now includes. It was assumed that students would bring fresh minds to an established science like chemistry, so uncluttered by conceptions about its content. Thus, they would be able to respond reasonably easily to the concreteness of chemistry's phenomena and then successfully engage with its more abstract conceptual explanations and principles by the stage of secondary schooling when their serious study of the subject began.

Alternative Conceptions Research

Many of the research studies of alternative conceptions about chemical phenomena and concepts have been done with students of ages 12–15, before or early in their study of chemistry. Sufficient has, however, been done on older secondary students, undergraduates, chemistry students, graduate teachers, and adults more generally, to support the claim that these conceptions are held by significant proportions of the respondents and are not a function of a particular age group. Indeed, they persist among substantial numbers of students and graduates who have studied chemistry extensively. Traditional teaching, testing and examining in chemistry often does not challenge these conceptions and students can hold them and still be quite successful in the usual tests and examinations.

Accordingly, the relevant findings are presented rather than details of the studies and of their respondents. These findings are more relevant in the cases of the A and C approaches to how they should be taught than they are to whether one approach is to be preferred to the other. In the case of the B approach the findings suggest this beginning to chemistry is unwise.

A. Conceptions of substances

A number of studies have reported that students can hold an elementary atoms conception in relation to the formation of a compound and a substance particle one in relation to its subdivision. Thus water is formed by the combination of atoms of hydrogen with atoms of oxygen, and rust is formed from atoms of iron, oxygen and hydrogen. When, however, macroscopic water or rust is subdivided — a process which is commonly seen as limited by the divider and not by what is being divided — the ultimate particles are 'water' and 'rust' *particles* or *atoms*.

Conversely, the common statement by teachers that 'water consists of hydrogen and oxygen' means to some students that a water molecule is formed by a major rearrangement of the nuclei and electrons from two hydrogen atoms and one oxygen atom. Many other students, however, ascribe to the same statement the meaning that water is a mixture of hydrogen and oxygen. This view is reinforced by their experience in elementary chemistry that this 'mixture' can be disentangled by electrolysis involving simple batteries. Some of them project this view on to what seems like the much more energetic process of boiling, by describing the bubbles of first air and then steam, as bubbles of hydrogen and oxygen. The distinction between pure substances and mixtures — a basic conceptual assumption in the substances approach — is thus quite confused even in this most simple of substances.

The distinction is similarly confused by students in their descriptions of other substances they encounter in school chemistry. In everyday life most materials are present as mixtures of substances but they are often treated as if they are pure. For example, water which has been rendered potable by the addition of appropriate substances is described as 'pure'. Newspaper headlines, like 'Chemicals found in food' or those that ignore other constituents like 'Vast deposits of uranium found' reinforce alternative conceptions that blend and overlap the ideas of pure substances and mixtures rather than distinguish them.

Mitchell and Gunstone (1984) found that the neat associations and distinctions the substances approach seeks to draw between *elements* with *atoms* as ultimate entities and *compounds* with *molecules* are often confused. This is not surprising when they are so soon not sustained in the descriptions of such simple substances as oxygen, nitrogen and chlorine.

The substances approach distinguishes between physical and chemical changes. Studies of students' conceptions of the phenomenon of dissolving common substances like salt or sugar in water reveal no such neat distinction. Firstly, 'dissolving' and 'melting' are commonly used by respondents rather indiscriminantly to describe this phenomenon, as they also are when melting is actually occurring. The difference between dissolving and melting is particularly difficult for students whose first language does not distinguish between *melt* and *dissolve* but it is by no means confined to these students. The use of heat in both melting and in accelerating solution in practice adds to the confusion. Secondly, the phenomena are not, in fact, distinct. Heating washing soda leads not to melting as it appears to, but to sodium and sulphate ions dissolving in the solid's water of crystallization! If sugar dissolves in water by dispersion of its constituent molecules, does salt dissolve or react? The hydrated cation and anion in a salt solution are not the cation and anion in the solid salt. Hydrogen chloride gas reacts with water as it dissolves to form ionic species that were not present in the original gas. Carbon dioxide dissolves and reacts with water. Metals like magnesium dissolve in acids, but in this case it is clearly a reaction process because a quite new substance, gaseous hydrogen, is formed. Like so many of the categorizations of nature that are used in science, physical and chemical changes are the ends of continuum of changes that can be, and often are perceived and conceived by beginning students in quite alternative ways to those intended.

B. Conceptions of atomic structure

Cros *et al.* (1986) found that 95 per cent of a large sample of university students did know about the fundamental particles in a simple Bohr model of atoms. Most of these, however, held alternative conceptions of the interactions between the constituents. For molecules and crystals the descriptions of the fundamental constituents and their interactions became more and more vague. Mitchell and Gunstone (1984), likewise, found that senior secondary students were commonly confused about the ideas of atoms and molecules in relation to elements and compounds. It is not that learners at almost any age do not have a vocabulary of atoms and molecules. They use these terms relatively easily, but with meanings that are far from those they now have in chemistry and physics.

Nussbaum (1985) began to illuminate the conceptions that lay behind these confused and vague descriptions of matter on an atomic scale. Despite students' use, in answering questions, of the phrase, *atoms as fundamental particles of matter*, a number of them did not have a discontinuous view of matter. Some saw matter as completely continuous, filling the space it occupies in each state of matter with varying degrees of spread. Others had a sense of atoms as particles, but embedded in a continuous form of the same substance. Another conception involved particles but particles aggregated together so that there was again a continuity.

Renström, Andersson and Marton (1990) develop these conceptions still further in an elegant series of studies in which Renström pursued students' ideas of particulate matter in a number of different contexts involving the sub-division of matter by mechanical means, by solution, and by change of state. The conceptions that described a macroscopic substance like salt or sugar as an aggregate of particles in some cases involved *elementary atoms* as the particles, but in others *substance particles* were involved. The *elementary atoms* in the former descriptions, furthermore, were quite often conceived in ways that were not very like the scientific view. Similarly, German students in describing crystal structure diagrams reported a view of atoms as having an exterior form like a skin and inside a more finely divided form. Thus the skin of an aluminium atom is like aluminium foil and inside there is a powdery aggregation of smaller aluminium atoms! Another conception of the structure of atoms is that they have a shell and nucleus, a reproduction in miniature of the surface and the interior these persons conceive crystalline substances in chemistry to have. That is, there is a shell and a nucleus, an inside and an outside. Like an egg, the inside of which is different in colour and texture to the outer shell, so the inside of white crystals of salt may be different and be where properties like taste are located, or as de Vos and Verdonk (1985) reported for lead nitrate and potassium iodide crystals, where the yellow substance resides that appears when these white crystals are ground together.

C. Conceptions of chemical reactions

Andersson and his colleague Renström carried out a number of studies of students' conceptions of chemical change. In 1986 he summarized their findings and a number that had been reported by others. Chemical changes were conceived in five ways. These are (i) 'just like that'; (ii) displacement; (iii) modification; (iv) transmutation; and (v) chemical interaction. The second, third and fourth were, furthermore, applied by the students to both the macroscopic world and to the world of atoms. Only about 20 per cent of students gave responses involving the sense of an interaction between the reactant chemicals that characterizes the fifth conception. A basic aspect of the first four alternative conceptions is the sense that the appearance of the product as a new substance is the result of a separate change in one or more of the original substances. The substances may mix with each other, but the new substances are not formed by chemical rearrangements between them. These findings parallel the way in which some students see chemical equations as quite independent chemical statements rather than as symbolic and quantitative representations of actual chemical reactions. Balancing equations is thus learning a set of rules. It is not, even among many of those mastering its rules, seen as a guarantee that the matter (in atomic terms) is conserved when new molecular substances are formed. Conservation of atoms or mass in chemical reactions is by no means axiomatic to many students due to their conceptions of what the new substances formed in reactions consist.

The role of energy in reaction systems has also been reported as a major source of these alternative conceptions. de Vos and Verdonk (1985) point out that it is the energy change in many examples of reactions that holds the focus of students' attentions *not* the new substances produced. The pop of hydrogen burning, the blinding flash of burning magnesium, etc. encourage many students to happily convert energy into matter and vice versa (Anderson and Renström 1983; Fensham, 1984) rather than to conserve the atoms of the matter involved.

This tendency, for attention to be deflected from the formation of a new substance as the essence of a chemical reaction, applies particularly to the cases of combustion and thermal decomposition reactions that have been so favoured (see page 16) as the exemplars to introduce a Reactions Approach. Indeed, as Andersson (1986) has amply illustrated, when heat is associated with a reaction, many students see it as a component that interacts with one of the original substances rather than as a condition promoting their mutual interaction.

Conceptions of States of Matter

Many findings have been reported of how students view solids, liquids and gases and the transitions between them (see for example, Stavy, chapter 16 in this volume). Alternative conceptions of each of the states and of their particular nature are common. Here only those associated with the gaseous state will be indicated because it is that state (as was suggested on page 17 far more than the solid or liquid states), that is important for teaching in introductions to chemistry.

A very widely held conception is that a gas is not a kind of substance at all. That is, gases do not have the usual associations of substances such as mass and its consequence of weight. Descriptions of gases by students can be of a smeared out continuum, or particulate but with particles that are much more densely contained than is the case in reality (Scott, 1987). This latter view is often reinforced by the diagrams in many text books of solids, liquids and gases as particles. A common conception is that a gas is air or some form of air. Boiling water is air leaving the water rather than liquid water changing to a gaseous form. The loss of water was liquid water carried out by large bubbles (Andersson and Renström, 1979). The gaseous form of water was perceived to be the tiny liquid water droplets formed as steam condenses. West, Fensham and Garrard (1982) found that undergraduate chemistry students often saw the evaporation of water below its boiling point as a property unique to water and not a change of state that generally occurs with many other liquid substances. The smell from liquids is confused with air contaminated by some process with the liquid substance. Nussbaum (1985) found that gas particles are thought to expand on heating like other substances rather than move faster.

The endowing of the macroscopic properties of a substance on to its atoms or molecules is a very common feature of many alternative conceptions. This holds for their actual properties. For example, copper atoms are thought to be malleable, and chlorine atoms are green, etc.

The scientific view that there is nothing between the atomic or molecular particles of a gas is particularly difficult for learners (Novick and Nussbaum,

1978). Brook, Briggs and Driver (1984) report that students are unable to use particle ideas to explain gaseous phenomena, but can provide macroscopic ones.

Constructivist Teaching of Chemistry

During the latter half of the 1980s, a number of teaching responses to these findings about students' alternative conceptions began to be reported. These have usually been associated with a generally constructivist view of learning. This means (i) that the existing views of the learners, that is their alternative conceptions, need to be taken very seriously; and (ii) that teaching becomes a set of events in which the learners are encouraged to engage their own minds actively with the observation of phenomena, and with the descriptions and explanations their peers and their teachers provide for them. It is through this active engagement of minds that each student's existing conceptions (and long-term memory structures) are believed to be modified, developed and generally reconstructed. The phrase 'Science Education: A Minds-On Approach', the title of a recent book by Duckworth, Easley, Hawkins and Henriques (1990) is an excellent short statement of what is intended.

In what follows some reports of such constructivist teaching relating to the different beginning approaches to chemistry are briefly described. These teaching initiatives often deliberately set out to link topics and concepts which more traditionally have been kept apart. It is not, therefore, really fair to report them under restricted headings but it does help to relate them to the earlier discussion.

A. Substances

Laverty and McGarvey (1991) reported on teaching the ideas of chemical substances involved in the formation of magnesium oxide from the combustion of a given mass of magnesium in air, and the formation of copper oxide by heating copper carbonate. They used a number of strategies that encourage constructivist learning including the teaching sequence of the Children's Learning in Science project (CLIS, 1987), metacognitive diaries, and group and class discussions. The practical work involved the students in oral and written reporting of their observations and their explanations of them.

The assessment, seen very much as also part of the teaching included individual interviews, homework tests and written tests in class which concentrated on questions and items that probed the students' macro- and micro- conceptual understanding of the substances involved. Assessment of this teaching approach found that the learners had made good learning gains on particulate (not atomic) representations of the substances and that they could use these representations to reveal and improve their understanding of elements, mixtures, and compounds.

B. Atomic structure and states of matter

Few attempts have yet been made to explore constructivist teaching of the Atomic Structure approach, but a number have been tried for States of Matter and the Kinetic Theory.

The conceptual studies have revealed daunting problems about the discontinuous view of matter and about particles on a molecular scale. These constitute a serious deterrent to any more ambitious ideas about student understanding of the structural details of substances at the atomic or molecular levels.

The Children's Learning in Science Project (CLIS, 1987) used a sequence that moves through six stages: A. orientation and elicitation of pupils' ideas; B. the nature of scientific theory and theory making; C. a pattern of properties of solids, liquids and gases; D. pupil's theory making; E. review, reflection and movement towards accepted theory; F. application and testing of accepted theory. This sequence covered thirteen lessons in which a number of the usual phenomena, particularly those involving gases, were introduced. They became the foci for exploring the pupils' ideas, for the sharing and interaction of these ideas, and for the teacher to pose situations that could challenge these ideas and, as new experiences, may lead to modification of the pupils' current views.

Scott (1987) and Johnston (1990), two of the schemes' developers, have very honestly reported their assessments of the scheme. A very positive outcome was that learners with a wide range of prior achievement could, with encouragement, participate actively in such a pedagogical approach and make significant conceptual learning gains. On the other hand, if conceptual change to the accepted theoretical position of contemporary school chemistry was the yardstick, a more conservative assessment was needed 'Rather than conceptual change there appeared to be a parallel development of particle ideas alongside already existing ones' (Scott). 'A clearer picture of conceptual problems about matter is emerging. Design and evaluation of teaching/learning strategies that are targeted at these specific conceptual problems is needed' (Johnston).

In a number of studies emphasis has been put on explicit recognition of the macro and micro levels as distinct and different perspectives for the same phenomena. Berkheimer *et al.* (1988) have prepared a teaching programme for Properties of Matter that for each phenomenon addresses these two levels as needing their own sets of conceptions. Unless the conceptions at the macro-level are dealt with, they argue, there is little point in moving to the micro-level. Conversely, however, it appears that students make little sense of phenomena like diffusion unless they know about atoms and their kinetic properties. Ten Voorde (1990) suggested, moreover, that there is a need to be quite selective in the phenomena that are chosen for this dual level approach. *Which macroscopic phenomena make it necessary to speak of microscopic particles*? He concludes that empirical chemical contexts alone do not usually necessitate the conclusion that substances have a particulate structure. Empirical physical contexts are more likely to require, and hence to lead to such a conclusion, as Ben-Zvi, Silberstein and Mamlok (1990) reported when teaching about small numbers of molecular species interacting energetically.

C. Reactions

Dierks (1989) developed a teaching sequence to chemical reactions in which students are encouraged to become aware of the fact that it is not unusual for a variety of questions to be asked about a chemical reaction. *What substances have disappeared? What characterizes the substances that have been formed? Is it a process that produces things with different characteristics? Is it a process in which the names of the products indicate the substances from which they are formed? What energy changes are associated?* The use of such a wide range of questions helped to bring out the variety of conceptions that the students held about the reaction. He goes on to stress the importance of using consistent language throughout these discussions about reactions. The reason for this is associated with the now well-established conceptual problems that students have in distinguishing the discontinous and continuous views that are used in chemistry. In introductory teaching he suggests that prose should be systematically used for continuous meaning, and that chemical symbols should be restricted to the discontinuous meanings when atoms and molecules are involved. *When hydrogen and oxygen are burned, water is produced. The atoms in the molecules of* H_2 *and in the molecules of* O_2 *are rearranged to form an* H_2O *molecule.* Mitchell and Gunstone (1984) have also reported considerable success with teaching that establishes the meanings that chemical equations embody about reactions by involving the rearrangement of small numbers of the atoms or molecules to give the new molecules before moving to the macroscopic level in which moles of the substance are involved.

Dierk's pedagogy draws heavily on the importance of learners recognizing the construct systems of others rather than that they should all have the same construct system. Such recognition was one of the new objectives for science education Fensham (1983) had suggested, perhaps anticipating the current interest in learning as conceptual addition rather than simply as conceptual change. The social process of communicating effectively with peers thus becomes an important pedagogy for teachers to facilitate alongside those that help each learner endeavour to construct meaning from the teacher's statements.

A constructivist position has also led de Vos and Verdonk (1985) to propose new approaches to the teaching of reactions which they see as the key entry point for chemistry. In a series of papers in 1985, 1986 and 1987 they outline their approach. In these they give some indication of the manner in which many traditional chemical experiments and their use in classrooms can be redesigned to facilitate the construction of meaning by the students. Their version of the potassium iodide/lead nitrate experiment referred to earlier is a case in point. These authors report that this approach to teaching chemical reactions has been successful in moving students from the conceptual alternatives that Andersson reported as so common (see page 20) to the one that involves interaction between the reactants and a reorganization of their atoms.

In 1989/90 de Vos published an introductory text book, *Chemistry in a Thousand Questions*, the title of which indicates how he has been very influenced by the spontaneous ideas about reactions and chemical phenomena that learners hold and by their questions which so often reflect the alternative conceptions they bring to each reaction phenomenon.

Conclusion

In the light of the conceptual findings about gases referred to earlier, and of the sober assessments of serious constructivist attempts to teaching kinetic theory via States of Matter, these topics must now be regarded as so difficult to learn that they are not only unrewarding early in the teaching of chemistry, but also a hindrance to the development of appropriate conceptions in later topics.

Millar (1990) made the interesting suggestion that the kinetic theory is difficult to teach because it is almost of no use to learners. By 'no use' he meant both the operational or functional sense of the words and also the cognitive/affective sense that could be associated with an improved understanding of matters of interest. He went on to argue for a much more oblique approach to the teaching of kinetic theory that quite gradually builds from the students' naive questions about matter, and their accumulation of examples and specific instances of its behaviour. The vague and tentative versions of what matter is like that Scott and Renström have reported may well be necessary steps for most students along this oblique pathway of learning. Such an approach contrasts starkly with the usual direct and early attempt to teach it via States of Matter, but there could be much gain if students did not construct firm alternative and unhelpful conceptions about atoms and gases so early.

To explore the consequences of postponing these topics, I examined the dozen or so chapters (topics) that follow in a number of introductory chemistry texts which locate States of Matter as one of their early topics. Almost without exception, none of the content in these subsequent chapters made reference to the phenomena and kinetic theory dealt with in the States of Matter, beyond the phenomenological assumption that solids, liquids and gases are obviously and recognizably different macroscopically.

Next, I considered the centrality of the idea in chemistry that matter consists of substances with characteristic and unique properties (see the opening definition on page 14) and the historical recognition and identification of these substances. By the seventeenth century when Boyle, Hooke and others were establishing quantitative relations that applied to air as a gas, no pure substances as gases had been identified, although a number of liquid and solid substances were known. Gas was equal to air. There were sulfurous air, smoke, marsh gas, and numerous foul and sweet smells, but they were all conceived of essentially as forms of air. Gases as distinct and distinguishable substances under normal temperatures and pressures took another century and genius like that of Scheele, Priestley and Lavoisier to be recognized. By the time that oxygen, nitrogen, sulphur dioxide and carbon dioxide had been identified, the number of solid and liquid substances had advanced several further orders of magnitude.

The very persuasive nature of the conceptions associated with phlogiston that had been popularized by Stahl in 1723 were, furthermore, a major inhibition to the recognition of gases. Hydrogen chloride was dephlogisticated muriatic acid and sulfur dioxide was phlogisticated vitriolic acid. This conceptual mixing of heat (energy) with substances to form new ones has an echo in the constructions of many learners today (see page 21).

Gases, historically, thus took much longer than solids and liquids to be recognized by chemists. Now, because of the apparent simplicity of gases in a molecular kinetic sense, and because their macroscopic response to pressure and temperature approximate to the simple and elegant Ideal Gas Equation, it has been too readily assumed, and become traditional, that the gaseous state of matter will be most readily comprehended by beginning learners of chemistry.

One of the recurring findings in the research studies is the way students give macroscopic explanations for chemical phenomena. This occurs even when they have had extensive teaching and learning of the kinetic molecular theory and of atomic and molecular structure. It is as if their knowledge of matter on an atomic scale is kept in isolation from chemistry as they experience it in the laboratory or learn about it in its macroscopic, phemonenal form. Whatever knowledge they have of the atomic scale of matter is not used as 'explanatory' of, or as a way of describing the behaviour of substances at large.

Recently there has been a very strong international interest in developing the curricula of the school sciences like chemistry so that they can be studied by much wider cross sections of students than has been the case hitherto. Most of the recent curricular reforms have drawn on the ideas of the Science — Technology — Society movement in order to gain the potential to attract and service these new target groups (see Fensham, 1991). The Salters chemistry course in England is typical of these developments and its authors state that this course 'starts with material and phemonena familiar to 13–16-year-olds from their own experiences or from TV, books, etc. Chemical concepts and explanations are introduced when they are needed in working on these everyday things. Industrial, technological, economic and social implications of chemistry are central to its study'.

An introductory study of chemistry in which concepts are given meaning by contexts that make sense to learners should have advantages for teachers wishing to take a constructivist approach. It should more easily enable connections to be made with the conceptions learners already hold, and students should thus be able to make the constructive links and reconstruction of their ideas more easily. Either of the Substance or Chemical Reactions approaches should fit with this association with the lives of the students outside of school. These curricular developments will not be of assistance to the Atomic Structure approaches. The phenomena of everyday and applied chemistry have little or no need for atomic scale or kinetic molecular explanations. They are described and explained in macroscopic terms and furthermore the substances involved have a continuous rather than a discontinuous sense. Thus, the students' own natural or intuitive senses of the continuity of matter will be reinforced rather than challenged.

References

Andersson, B. (1986). Pupils' explanations of some aspects of chemical reactions. *Science Education. 70*, pp. 549–63.

Andersson, B. (1990). Pupils' conceptions of matter and its transformations (age 12–16). *Studies in Science Education, 18*, pp. 53–85.

Andersson, B. and Renström, L. (1979). *Heat and Temperature: Boiling*, Mölndal, Sweden: Department of Education, University of Gothenburg.

Andersson, B. and Renström, L. (1983). *Boiling in a Closed System. The Freon Experiment.* Mölndal, Sweden: Unpublished report for Department of Education, University of Gothenburg.

Bent, H. (1981). Making things easy and making things difficult. In *Conference Proceedings, Sixth International Conference on Chemical Education, University of Maryland*, College Park, MD: University of Maryland, pp. 91–104.

Ben-Zvi, R., Silberstein, J. and Mamlok, R. (1990). Macro-micro relationships: a key to the world of chemistry. In Lijnse, P.L. Licht, P. de Vos, W. and Waarlo, A.J. (Eds.). *Relating Macroscopic Phenomena to Microscopic Particles.* Proceedings of a seminar at Utrecht Centre for Science and Mathematics Education, University of Utrecht, pp. 183–97.

Berkheimer, G.D. *et al.* (1988). *Matter and Molecules, Teacher's Guide: Science Book.* Occasional paper No. 121. East Lansing, MI: Institute for Research on Teaching, Michigan State University.

Bouma, H., Sutton, C. and Brandt, L. (1988). *Words as Tools in Science Lessons.* Leicester: School of Education, University of Leicester; — see also Sutton, C.R. (1992). *Words, Science and Learning.* Milton Keynes: Open University Press.

Brook, A., Briggs, H. and Driver, R. (1984). *Aspects of Secondary Students' Understanding of the Particulate Theory of Matter*. Leeds: Centre for Studies in Science and Mathematics Education, University of Leeds.

Bucat, R.B. and Cole, A.R.H. (1988). The Australian Academy of Science School Chemistry Project. *Journal of Chemical Education*, *65*, pp. 777–9.

CLIS (1987). *Approaches to Teaching the Particulate Theory of Matter*. Children's Learning in Science Project, Leeds: Centre for Studies in Science and Mathematics, University of Leeds.

Cros, D. *et al.* (1986). Conception of first year university students of the constituents of matter and of notions of acids and bases. *European Journal of Science Education*, *8*, pp. 305–14.

Dierks, W. (1989). An approach to the educational problem of introducing the discontinuum concept in secondary chemistry teaching and an attempted solution. In Lijnse, P.L., Licht, P., de Vos W. and Waarlo, A.J. (Eds.). *Relating macroscopic phenomena to microscopic particles*. Proceedings of a seminar at Centre for Science and Mathematics Education, University of Utrecht, pp. 177–82.

de Vos, W. and Verdonk, A.H. (1985, 1986 and 1987). A new road to reactions, parts 1–5. *Journal Chemical Education*, *62*, pp. 238–40, and 648–9, *63*, pp. 972–4, *64*, pp. 692–4 and 1010–3.

Duckworth, E., Easley, J., Hawkins, D. and Henriques, A. (1990). *Science Education: A Minds-on Approach for the Elementary Years*. Hillsdale, NJ: Lawrence Erlbaum Associates.

Fensham, P.J. (1983). A research base for new objectives of science teaching. *Science Education*, *67*, pp. 3–12.

Fensham, P.J. (1984). Conceptions, misconceptions and alternative frameworks in chemical education. *Chemical Society Reviews*, *13*, pp. 199–217.

Fensham, P.J. (1988). Approaches to the teaching of STS in science education. *International Journal of Science Education*, *10*, pp. 346–56.

Fensham, P.J. (1991). Science and Technology. In Jackson, P.W. (Ed.). *Handbook for Research on Curriculum*. New York: Macmillan, AERA, pp. 789–829.

Gillespie, R.J. (1976). Chemistry — fact or fiction? Some reflections on the teaching of chemistry. *Chemistry Canada*, *28*, pp. 23–8.

Johnston, K. (1990). Students' responses to an active learning approach to teaching the particulate theory of matter. In Lijnse, P.L. Licht, P. de Vos, W. and Waarlo, A.J. (Eds.). *Relating Macroscopic Phenomena to Microscopic Particles*. Proceedings of a seminar at Centre for Science and Mathematics Education, University of Utrecht, pp. 247–65.

Lagowski, J.J. (1985). What happened to descriptive chemistry? *Journal of Chemical Education*, *62*, p. 915.

Laverty, D.T. and McGarvey, J.E.B. (1991). A 'constructivist' approach to learning. *Education in Chemistry*, *28*(4), pp. 99–102.

Millar, R. (1990). Making sense: what use are particle ideas to children? In Lijnse, P.L. Licht, P. de Vos, W. and Waarlo, A.J. (Eds.). *Relating Macroscopic Phenomena to Microscopic Particles*. Proceedings of a seminar at Centre for Science and Mathematics Education, University of Utrecht, pp. 283–93.

Mitchell, I.J. and Gunstone, R.F. (1984). Some student conceptions brought to the study of stoichiometry. *Research in Science Education*, 14, pp. 78–88.

Novick, S. and Nussbaum, J. (1978). Junior high school pupils' understanding of the particulate nature of matter: An interview study. *Science Education*, *62*, pp. 273–81.

Nussbaum, J. (1985). The particulate nature of matter in the gaseous state. In Driver, R. Guesne, E. and Tibergien, A. (Eds.). *Children's Ideas in Science*. Milton Keynes: Open University Press, pp. 124–44.

Renström, L., Andersson, B. and Marton, F. (1990). Students' conceptions of matter. *Journal of Education Psychology*, *82*, pp. 555–69.

SATCHELL, D.P.N. (1982). Beginning O-level chemistry. *Chemistry in Britain, 18*, p. 161.

SCOTT, P. (1987). The process of conceptual change in science: a case study of the development of a secondary pupil's ideas relating to matter. *Proceedings of Second International Seminar on Misconceptions and Educational Strategies in Science and Mathematics*, Vol II, Ithaca, NY: Cornell University, pp. 404–19.

STAVY, R. (1985). Children's ideas about liquids and solids. *European Journal of Science Education, 1*, pp. 407–21.

STAVY, R. (1988). Children's conceptions of gases. *International Journal of Science Education, 10*, pp. 553–60.

TEN VOORDE, H.H. (1990). On teaching and learning about atoms and molecules from a Van Hiele point of view. In LIJNSE, P.L. LICHT, P. DE VOS, W. and WAARLO, A.J. (Eds.) *Relating Macroscopic Phenomena to Microscopic Particles*. Proceedings of a seminar at Centre for Science and Mathematics Education, University of Utrecht, pp. 81–104.

WEST, L.H.T., FENSHAM, P.J. and GARRARD, J.E. (1982). *Describing the Cognitive Structure of Undergraduate Chemistry Students*. Canberra: Education Research and Development Committee.

Chapter 3

Generative Science Teaching

Merlin C. Wittrock

For many years science teaching has focused on ways to engage learners' generative thought processes in the learning of scientific and mathematical concepts and principles that will transfer to facilitate related learning in science and to enhance problem-solving in everyday situations. One of the more intriguing parts of research and thought about science teaching has been its ambition since antiquity to train intelligence by engaging the learners' generative thought processes in the construction of meaning for concepts and principles that will transfer and will solve problems. For example, Plato taught a slave boy the Pythagorean theorem as a way to train intelligence, or virtue as it was then called. Charles Judd, in the first decade of this century, taught boys the principle of refraction of light as a way to increased transfer of ability to solve a practical problem of hitting a target submerged under different depths of water.

Modern day research on cognition in science teaching focuses upon these same generative thought processes of learners that have intrigued science teachers since antiquity. These thought processes include the learners' knowledge and experience, comprehension strategies, and metacognitive strategies. Many of the modern day research studies show the critical importance for science teaching of the learners' preconceptions in electricity (Osborne and Wittrock, 1983 and 1985), composition of matter (Renström, Andersson and Marton, 1990), and gravity and cosmology (Nussbaum and Sharon-Dagan, 1983). Other recent studies show the critical importance for science learning of the students' comprehension strategies and metacognitive processes. For example, more advanced learners, compared with beginning students, produce better representations of problems (Peled and Wittrock, 1990) and classify problems better according to their underlying scientific or mathematical principles (Alexander and Judy, 1988).

The more advanced science learners also have better developed self-regulatory and self-monitoring metacognitive skills (Glaser, 1990), that facilitate planning and use of procedural knowledge as well as declarative knowledge. These more competent science learners also better transfer their science learning than do their less competent counterparts. Lehman, Lempert and Nisbett (1988) studied the transfer effects of graduate training in psychology, law, medicine, and chemistry. Training in psychology and medicine increased ability to transfer and use scientific reasoning in the solution of everyday clinical problems. Training in law increased ability to reason logically about everyday, complex legal cases and problems. Training in chemistry, unlike the training in psychology, medicine and law that emphasized transfer of scientific or logical methods to practical contexts, did not facilitate ability to solve everyday problems in chemistry. In addition to domain or content-specific learning at the graduate school level, it is important to engage learners' generative thought processes in the representation, transfer, application and self-control of knowledge and action.

With younger students, including mentally retarded children, related findings occur. Belmont, Butterfield and Ferretti (1982) found that only seven of 114 studies of instruction with young mentally retarded children produced transfer. In six of these seven studies that produced transfer, the children were taught higher-order thinking skills, especially self-management skills. The training of metacognitive thought processes apparently increases productive thinking and increases problem solving ability.

Related findings occur in research on science teaching at the high school level. Baird (1986) studied metacognitive learning in ninth grade general science and in eleventh grade biology classes. He used metacognitive strategies similar to those used to train attention and reading comprehension (for example, What is the topic? Why am I doing it? Do I fully understand this?). Comprehension and understanding of science improved in this realistic context, although these teaching practices contrasted with the more passive normal procedures used by the teachers of these science classes.

White and Gunstone (1989) show the importance of metacognition in conceptual change. From the work of the Project to Enhance Effective Learning (PEEL), which teaches metacognition in secondary school science, they argue that metalearning plays an important role in facilitating conceptual change in science classes.

In sum, the recent cognitive research on science teaching shows the importance of several levels of learning: (i) metacognitive learning; (ii) learning strategies; (iii) executive skills; (iv) subject matter learning; and (v) knowledge of principles, rules, and specific situated knowledge.

Metacognitive learning, that is, awareness and control over one's thought processes during learning, should be included in comprehensive conceptions of science teaching. Metacognition increases transfer among students of different abilities and across different subject matters, including science. Metacognitive strategies can be taught in science classrooms (for example, Baird, 1986). These strategies increase the transfer of science learning; and they facilitate conceptual change.

With the difficult problems of teaching science to all students, these findings about metacognition in conceptual change offer significant promise. Science for all students (Fensham, 1985) involves fundamental revisions in the way we think about science learning and science teaching. The learners' alternative conceptions of phenomena in science and the learners' thought processes, including metacognition and strategies of comprehension, attention, attribution, and generation, become especially important in teaching science for all students. Conventional methods of covering the subject matter and presenting only the scientists' view of scientific phenomena clearly do not effectively teach science to all students.

One way to try to improve upon current and common practices begins with the development of a conception of science learning as the learners' generation of meaning for scientific events through changing their alternative conceptions.

Within that framework, science teaching involves (i) learning about the students' alternative conceptions, beliefs, attributions, and related cognitive and affective thought processes; and (ii) teaching students to use their knowledge, beliefs, and metacognitive and affective thought processes to generate new, fruitful, and transferable conceptions that have personal and everyday meaning and significance. In the following section, I want to present a model of science teaching that includes but goes beyond student preconceptions to consider motivation, attention, generation and metacognition. The model presented in the following section builds upon earlier, related work on these problems (Wittrock, 1974a, 1990 and 1991; Osborne and Wittrock, 1983 and 1985).

A Model of Generative Science Teaching

Effective teaching of science to all students involves the construction of a comprehensive model of the cognitive processes involved in learning science from instruction. These processes include the learner's background knowledge and alternative conceptions, but also the sometimes neglected processes of attention, motivation and attribution, generation and metacognition.

Structural models of the storage of information, which include structural schema theories, focus on the memory structures that represent knowledge, such as schemata and scripts. Although these structural models inform us well about the nature of stored information, they inform us less well about the acquisition or learning of that stored and structured information, for example, of the generative processes involved in relating new information to knowledge and experience. Neither do structural schema theories focus upon the central teaching problems of facilitating learning from instruction, such as learning science from classroom teaching, including the wide range of motives, interests, conceptions, and abilities found in typical school settings.

The model of generative teaching, described in the following paragraphs, differs from structural models of the storage of information in several ways. First, it is a functional model, not a structural model, as are several schema theories. As a functional model of learning its focus is upon the cognitive and the neural processes that learners use to comprehend and to understand science. The model states that these functions or processes of understanding science involve active learner generation of two types of meaningful relations. The first type of generated meaningful relation is between learner knowledge and experience (for example, alternative conceptions on the one hand, and new, or to-be-learned information and conceptions, on the other hand). The second type of generated meaningful relation is among the parts of the information to be learned. The model's conceptualization of understanding in science leads to testable predictions about the cognitive processes learners use to comprehend science.

Second, it is also a functional model of teaching, or more accurately, of learning from teaching. The model predicts that comprehension and understanding in science can be facilitated by teaching that leads learners to generate analogies, metaphors, problems, and related devices that build the two types of meaningful relations described in the previous paragraph. Again, the model leads to testable predictions, in this case about the teaching conditions that foster meaningful learning of science concepts and information.

Third, the model builds upon research that goes beyond generation as a process in conceptual change to include attention, motivation, attribution, and metacognition. In research on science learning and teaching these important processes have often been identified and discussed. For example, the intrinsic motivation of learning about scientific events and the attention-getting properties of unexpected results in demonstrations in science classes have often served as the bases for facilitating science teaching. More recently, the important role of learner awareness and self-control, that is metacognition, in the facilitation of understanding and transfer in science learning has been discussed (White and Gunstone, 1989; Wittrock, 1991). This functional model of generative science learning incorporates these processes in an attempt to facilitate learning from science teaching.

In sum, the model consists of four functional, cognitive processes: (i) knowledge, experience, and conceptions; (ii) motivation and attribution; (iii) generation; and (iv) metacognition.

Knowledge, Experience and Conceptions

In science learning the critical role of the learners' background knowledge and conceptions has been clearly established. For example, Osborne (1981) found that elementary school children in several English speaking countries believe that direct current in a simple electrical circuit, consisting of a battery, a bulb, and their connecting wires, flows in three ways: (i) from one side of the battery to the bulb; (ii) from each side of the battery to each side of the bulb; and (iii) from one side of the battery through the bulb to the other side of the battery, equal in amplitude through the whole circuit. These preconceptions firmly resist change. Demonstrations with ammeters that show that DC current does not flow as predicted by the first two models usually produce no change in the learner's conceptions. Instead, the students defend their models as true in the everyday world, and discredit the demonstrations as anomalous and true only in the classroom, a world controlled by powerful science teachers who create magical events.

Students' conceptions of the earth (Nussbaum, 1979; Nussbaum and Sharon-Dagan, 1983), of the particulate nature of gases (Benson, Wittrock and Baur, in press), and of gravity (Gunstone and White, 1981), as well as of other science concepts, also often show great resistance to change through instruction. The volume on conceptual change edited by West and Pines (1985) discusses these issues further.

These student conceptions represent a fundamental component of the model of generative science teaching. They represent the knowledge base for building relations between the concepts to be learned and experience summarized in alternative frameworks. The generation of meaningful, scientific conceptions clearly involves these often unscientific conceptions. The identification of these student conceptions implies an advance in the design of science teaching for all students. No longer can science teaching focus only on presenting the scientists' views of physical events, or on covering the subject matter of science. Science teaching also involves understanding the students' views of science concepts. Teaching involves more than showing students the incorrectness of their beliefs that work quite well for them everyday in realistic contexts. It involves more than setting up dissonances between students' models and teacher controlled demonstrations. It involves leading students to test and develop their models and thought processes in familiar contexts, which they believe are real, representative of everyday experience, and under their control rather than subject to manipulation by powerful people who cause clever but false things to happen.

Motivation

Generative science learning involves a distinctive type of motivation. The conscious generation of meaning from experience involves taking responsibility for learning, and believing that one can succeed at understanding complex everyday experiences through actively generating and testing concepts. The extensive research on attribution for learning, for example, Weiner (1979); on self-regulation and self-efficiency, for example, Pintrich and De Groot (1990); on perceived control, for example, Zimmerman and Martinez-Pons (1990); and on learned helplessness, for example, Dweck (1975) shows the importance of this component in learning from instruction. Although it is not commonly developed in models of science teaching, this distinctive type of motivation offers much promise for increasing generative science learning through active involvement of students in the construction and testing of concepts in science.

Attention

Recent neural research and cognitive research indicates that the brain is a model builder. It has short term and long term attentional mechanisms (Wittrock, 1980) that react to discrepant events that indicate one's model of an event may need repair or revision.

Questions and objectives direct attention. Problems, discrepant events, and enigmas, such as those commonly presented in science classes, direct attention, especially short-term attention. The direction of long-term voluntary attention toward the construction of a scientific conception of a discrepant event presents a challenge for generative learning. Students easily finesse discrepant events, rather than use them as occasions to construct meaningful relations between the event and their knowledge and experience that would lead to a revised and improved conception. Voluntary, sustained attention focused on the construction of meaningful explanations and useful scientific conceptions is a centrally important component of generative science training.

Generation

To learn science with understanding, students generate a model or an explanation that organizes information into a coherent structure, and that relates the information to their knowledge and experience. Comprehension of science concepts involves building two types of relations, (i) among the parts of the new information; and (ii) between the new information and the learner's knowledge and experience. This model of generative learning has been described and empirically tested in the teaching of science, mathematics, reading, and economics (for example, Wittrock, 1974a, 1974b, 1981, 1990 and 1991; Wittrock and Carter, 1975; Doctorow, Wittrock and Marks, 1978; Wittrock and Alesandrini, 1990; Linden and Wittrock, 1981; Osborne and Wittrock, 1983 and 1985; Peled and Wittrock, 1990). These and related studies show that generative learning processes play important roles in learning with understanding in these subjects; and that generative teaching procedures often facilitate learning with comprehension. For example, Doctorow *el al.* (1978) showed a doubling of comprehension among sixth graders when students generated summaries and headings for each paragraph of a text in a commercially published reader, compared with students who were given the same amount of time to read the same text without generating headings and summaries of each of its paragraphs. Linden and Wittrock (1981) increased reading comprehension among fourth and fifth grades by 75 per cent by asking them to generate interpretations and images that related the sentences of the stories to one another, and that related the stories to their knowledge and experience. Kourilsky and Wittrock (1988 and 1992) showed sizeable gains in the comprehension of economics among secondary school students taught by generative teaching procedures to construct these two kinds of relations mentioned previously. Teachers can use a variety of procedures to facilitate generative learning. First, familiar materials taken from the students' everyday experience, familiar words, and familiar believable contexts and problems facilitate generative learning. Teachers can use the following devices to facilitate student generation of relations between their background knowledge and believable, familiar contexts and materials: analogies; metaphors; images; diagrams; examples; demonstrations; pictures; paraphrases.

Teachers can use a variety of procedures to facilitate student generation of relations among the parts or the components of science concepts and science frameworks: titles; headings; questions; objectives; summaries; tables; problems; explanations.

Teachers can also lead students, as they become more proficient, to generate these types of relations. For example, students can learn to generate their own summaries or explanations as they work on a physics problem involving force and motion. Students can also be asked to generate their own summaries, metaphors, analogies, explanations, diagrams, concept maps, pictures, and fortune lines, as ways to facilitate generative learning.

Students can be given these relations and asked to do something mentally with them. Or, students can be asked to generate these relations. The choice between these different teaching procedures involves the students' background knowledge and chances of success at using it. It is what the students are asked to do mentally with these relation building teaching procedures, not only what the teacher does, that determines their effectiveness. Instruction can be direct and still involve generative learning, provided the students are led to examine, apply, test, or just think about the concepts. However, when the students have the appropriate background information, instruction other than direct instruction (i.e., telling them the concepts and the answers) often leads to enhanced interest, motivation, attention, and generation of meaningful relations that facilitate learning with understanding and a sense of competence that does not come as well by student learning from teacher telling.

Metacognition

Awareness and self-control over their own thought processes also can be taught to students. They can learn from teaching that different types of problems in science involve different structures, different heuristics, and different strategies for their solutions.

These metaprocesses of science learning seem especially important in a 'science for all students'. Their value has been shown in research on attention among fourth and fifth graders (Douglas, Parry, Martin and Garson, 1976) and in children's problem solving in elementary chemistry and physics, using children of a broad range of abilities (Swanson, 1990).

Students can also learn to monitor their own use of the generative learning strategies discussed in the previous section on generation. These individual strategies can be organized into sequences of strategies that facilitate comprehension. Wittrock and Kelly (1984) taught functionally illiterate young adults to organize and to monitor their generative reading comprehension strategies into a sequence of activities to perform (i) before, (ii) during, and (iii) after reading. This metacognitive strategy facilitated reading comprehension by 20 per cent, compared with a comparable group of randomly assigned students who were not taught this sequence of learning strategies nor a strategy for monitoring their use.

Students can also be taught a model of comprehension in science. What does it mean to learn science with understanding? How does learning science with understanding differ from learning science in a less meaningful fashion? Students need to understand the purpose and utility of learning science. They often profit from learning the utility and everyday value of science. When they learn how to transfer their classroom science learning into a meaningful understanding of their everyday world, that is, to generate relations between classroom science and everyday experience, their achievement and interest in science increases, and so may their feeling of competence in ability to learn science. These elements of a generative science teaching model relate directly to the problems of creating a learnable 'science for all students'.

Generative Teaching of Science

The model presented in this paper leads to a practical plan for teaching science. The elements of that plan are as follows:

Knowledge, experience, and conceptions

- Learn the students' conceptions and beliefs about science, beliefs about their own abilities to learn science, and beliefs about the meaning and usefulness of science in their daily lives.
- Learn the students' conceptions of what they must do to learn science.
- Teach the students that learning with understanding is a generative process. It is different from passively reading and remembering information.

Motivation

- Teach the students a distinctive type of motivation that involves:
  (a) taking responsibility for learning;

(b) believing that they can and will succeed at understanding science conceptions and through them gain a deeper understanding of complex everyday experience.

- Design instruction to enable students to experience frequent success at understanding science through using generative learning procedures.

Attention

- Teach students to attend to the central problems of constructing meaning for science concepts. Teach them to focus their attention on the problems of generating a structure of information and on the problems of relating science to other subjects and to everyday experience.

Generation

- After learning the students' models, beliefs, and conceptions, teach a scientific model in relation to the students' models. The primarily goal is *not* to 'cover the subject matter' or to 'present only the scientists' model' or to 'show that the students' models are wrong'. Instead, the primary goal is to lead the students to generate a more scientific and more useful model than they now have. That goal may not be attained quickly because it involves student generation of understanding. That generation of meaning requires student effort and thought that goes beyond learning from teacher reward and teacher punishment.
- Generate for them, initially at least, examples and applications of the two types of relations that facilitate comprehension: (i) relations among the concepts to be learned; and (ii) relations between these science concepts and student knowledge and experience.
- Lead students to generate these two types of relations, as a way to comprehend science.
- Design instruction to involve these two types of relations and to increase student ability to use the generative learning strategies that enhance comprehension through relation building.

Metacognition

- Teach students to become aware of and to consciously use their thought processes, including learning strategies, comprehension strategies, attention-directing strategies, attributions, plans, and monitoring strategies to comprehend, transfer, and apply science concepts to attain a greater understanding of their world and an enhanced ability to solve everyday problems.

References

Alexander, P.A. and Judy, J.E. (1988). The interaction of domain-specific and strategic knowledge in academic performance. *Review of Educational Research*, *58*, pp 375–404.

Baird, J.R. (1986). Improving learning through enhanced metacognition: A classroom study. *European Journal of Science Education*, *8*, pp. 263–82.

Belmont, J.M., Butterfield, E.C. and Ferretti, R.P. (1982). To secure transfer of training, instruct self-management skills. In Detterman, D.K. and Sternberg, R.J.

(Eds.). *How and How Much Can Intelligence be Increased.* Norwood, NJ: Ablex, pp. 147–54.

Benson, D.L., Wittrock, M.C. and Baur, M.E. (in press). Students' preconceptions of the nature of gases. *Journal of Research in Science Teaching.*

Cook, L.K. and Mayer, R.E. (1988). Teaching readers about the structure of scientific text. *Journal of Educational Psychology*, *80*, pp. 448–56.

Doctorow, M.J., Wittrock, M.C. and Marks, C.B. (1978). Generative processes in reading comprehension. *Journal of Educational Psychology*, *70*, pp. 109–18.

Douglas, V.I., Parry, P., Martin, P. and Garson, C. (1976). Assessment of a cognitive training program for hyperactive children. *Journal of Abnormal Child Psychology*, *4*, pp. 389–410.

Dweck, C. (1975). The role of expectations and attributions in the alleviation of learning helplessness. *Journal of Personality and Social Psychology*, *31*, pp. 674–85.

Fensham, P. (1985). *Science for all. Journal of Curriculum Studies*, *17*(4), pp. 415–35.

Glaser, R. (1990). The reemergence of learning theory within instructional research. *American Psychologist*, *45*, pp. 29–39.

Gunstone, R.F., and White, R.T. (1981). Understanding of gravity. *Science Education*, *65*, pp. 291–9.

Kourilsky, M. and Wittrock, M.C. (1988). Verbal and graphical strategies in the teaching of economics. *Teaching and Teacher Education*, *3*, pp. 1–12.

Kourilsky, M. and Wittrock, M.C. (1992). Generative teaching: An enhancement strategy for the learning of economics in co-operative groups. *American Educational Research Journal*, *29*(4), pp. 861–876.

Lehman, D.R., Lempert, R.O. and Nisbett, R.E. (1988). The effects of graduate training on reasoning. *American Psychologist*, *43*, pp. 431–42.

Linden, M. and Wittrock, M.C. (1981). The teaching of reading comprehension according to the model of generative learning. *Reading Research Quarterly*, *17*, pp. 44–57.

Nussbaum, J. (1979). Children's conceptions of the earth as cosmic body: A cross age case study. *Science Education*, *63*, pp. 83–93.

Nussbaum, J. and Sharon-Dagan, N. (1983). Changes in second grade children's preconceptions about the earth as a cosmic body resulting from a short series of audio-tutorial lessons. *Science Education*, *67*, pp. 99–114.

Osborne, R.J. (1981). Children's ideas about electric current. *New Zealand Science Teacher*, *29*, pp. 12–19.

Osborne, R.J. and Wittrock, M.C. (1983). Learning science: A generative process. *Science Education*, *67*(4), pp. 489–504.

Osborne, R.J. and Wittrock, M.C. (1985). The generative learning model and its implications for science education. *Studies in Science Education*, *12*, pp. 59–87.

Peled, Z. and Wittrock, M.C. (1990). Generated meanings in the comprehension of word problems in mathematics. *Instructional Science*, *19*, pp. 171–205.

Pintrich, P.R. and De Groot, E.V. (1990). Motivational and self-regulated learning components of classroom academic performance. *Journal of Educational Psychology*, *82*, pp. 33–40.

Renström, L., Andersson, B. and Marton, F. (1990). Students' conceptions of matter. *Journal of Educational Psychology*, *82*, pp. 555–69.

Swanson, H.L. (1990). Influence of metacognitive knowledge and aptitude on problem solving. *Journal of Educational Psychology*, *82*, pp. 306–14.

Weiner, B. (1979). A theory of motivation for some classroom experiences. *Journal of Educational Psychology*, *71*, pp. 3–25.

West, L.H.T. and Pines, A.L. (1985). *Cognitive Structure and Conceptual Change.* New York: Academic Press.

White, R.T. and Gunstone, R.F. (1989). Metalearning and conceptual change. *International Journal of Science Education*, *11*, pp. 577–86.

Wittrock, M.C. (1974a). A generative model of mathematics learning. *Journal for Research in Mathematics Education*, *5*, pp. 181–97.

Wittrock, M.C. (1974b). Learning as a generative process. *Educational Psychologist*, *11*, pp. 87–95.

Wittrock, M.C. (1980). Learning and the brain. In Wittrock, M.C. (Ed.). *The Brain and Psychology*, New York: Academic Press, pp. 371–403.

Wittrock, M.C. (1981). Reading comprehension. In Pirozzolo, F.J. and Wittrock, M.C. (Eds.). *Neuropsychological and Cognitive Processes in Reading*, New York: Academic Press, pp. 229–60.

Wittrock, M.C. (1990). Generative processes of comprehension. *Educational Psychologist*, *24*, pp. 345–76.

Wittrock, M.C. (1991). Generative teaching of comprehension. *Elementary School Journal*, *92*, pp. 167–82.

Wittrock, M.C. and Alesandrini, K. (1990). Generation of summaries and analogies and analytic and holistic abilities. *American Educational Research Journal*, *27*, pp. 489–502.

Wittrock, M.C. and Carter, J. (1975). Generative processing of hierarchically organized words. *American Journal of Psychology*, *88*, pp. 489–501.

Wittrock, M.C. and Kelly, R. (1984). *Teaching Reading Comprehension to Adults in Basic Skills Courses*. Three volumes. Final Report, University of California, Los Angeles, Graduate School of Education.

Zimmerman, B.J. and Martinez-Pons, M. (1990). Student differences in self-regulated learning: Relating grade, sex, and giftedness to self-efficacy and strategy use. *Journal of Educational Psychology*, *82*, pp. 51–9.

Chapter 4

Constructivism, Re-constructivism and Task-orientated Problem-solving

Mike Watts

In this chapter I review the strengths and limitations of a particular pedagogical strategy and illustrate this both in terms of explicit school science content and particular classroom practice. That strategy is called open-ended task-orientated problem solving, an expression I hope to unpick as the paper progresses.

Principally, though, I address two questions:

- How can task-orientated problem solving be used as a mechanism to promote constructivist conceptual change in a particular science content area?
- What makes this way of working *good constructivist* classroom practice?

With these two questions I frame the structure of the paper though the balance is tipped towards exploring the first in detail and the second in principle.

Actionable Knowledge

> Watch out Japan. Here comes Tracy Logan, Tracy Logan is a typical British sixteen year old, leaving school this year. But to Japan, and our other international competitors, she's a big threat.... That's because this year she'll be starting two years paid skill straining on the new YTS.... Tracy will be spending the next two years learning how to take trade away from them for a change. (MSC Advertisement, *Guardian*, 28 January 1986)

This advertisement for the Manpower Services Commission appeared in the British press in 1986. Given the state of trade today it was somewhat wildly optimistic. However, the most poignant part, in my view, is not governments' inability to capitalize on the talent of youth so much as the need to undertake 'skill training' with such fanfare in the years immediately after compulsory schooling. So, to begin this chapter I feel the need for basic questions: Why teach science in school? More importantly: Why should young people at school attempt to *learn* it? And how can their changing ideas be embedded in their daily lives so that they learn within a structured framework?

There are a number of (idealistic?) responses to these questions, adapted here from a description by the Association for Science Education (ASE, 1981):

- science is both deeply interesting and beautiful in construction. The pursuit of knowledge in science is an end in itself, an intellectual activity leading to the creation of further research and knowledge;
- science is part of the world of ideas, its history, philosophy, literature, pedagogy and social institutions contribute to the culture of our societies;
- science can be a passport to employment: high achievement in science can lead to a wide range of life chances, bringing understanding to use in industrial, commercial and social situations;
- science allows active participation in the process of democratic decision-making: understanding how some principles and laws in science are used can provide a basis for rational choice in life contexts;
- science brings understanding of physical phenomena in everyday life, so providing a creative basis for the pursuit of daily contexts and leisure activities.

While these may be over-romantic for many, I want to show that some small part of this can be delivered through 'task-orientated problem solving' in the school classroom. Most of us for most of the time are heavily constrained by syllabuses, curricula, timetables, assessment schedules, textbooks, school structures and so on. However, I believe we also need to explicate some of the aims and ambitions we have for the activities we develop — and in this case I err away from the ideal and towards the pragmatic, to focus on the latter three aims in my list. Behind these three lies the notion that scientific knowledge should be useful and lead to practical action — what Layton (1991) refers to as 'cognition in practice' and the construction of 'situated knowledge'.

In his article Layton proposed the following diagram (see figure 4.1) and suggests that there has been considerable research on the left hand side but little on the right. Here he is referring (on the left) to the wealth of research data

Figure 4.1: Construction and de-/re-construction of scientific knowledge

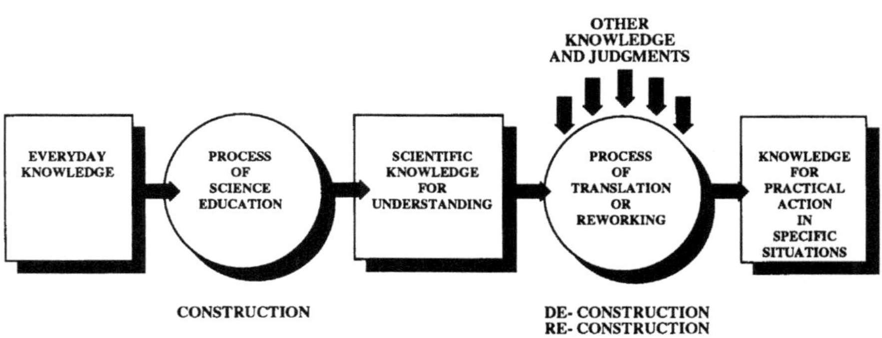

about learners' scientific construction and interpretation of phenomena in everyday life (see, for example, the bibliographies of Carmichael *et al.*, 1990, and Pfundt and Duit, 1991). A range of this research is mentioned in other parts of this book. In his article Layton goes on to say: 'Although much research has been conducted on problem solving, little of this has been in relation to technological and other problems involving practical action. We lack understanding of the process and its development characteristics.'

Science needs to be relevant to students' everyday lives since this real context provides the roots from which their studies *should* be drawn. It needs to be related to their hobbies and modern lifestyles; to current affairs and television news; to people and practices in the world; to ideas and creative thinking at work; to economic and industrial success. The movement for 'relevant' is not new and throughout the eighties in the United Kingdom it has helped to shape school science, so that schemes like *SATIS (Science and Technology in Society)* and *Science and Society* are motivated by the need to relate the 'application' of science to current issues in society. Such materials tend to be lively and interesting, though it is sometimes a struggle to keep the laws and principles of science within the bounds of everyday activity.

While some of this material is based upon project work, the outcomes of student activity tends to be 'software' — written, commonly numeric answers to textual problems. In this chapter I explore 'hardware' solutions where pupils construct and test some physical outcome to their ideas. So, to set the scene, in three broad examples of classroom and off-site work in action I examine the implementation of task-orientated problem solving in secondary school science. My examples are drawn from the work of:

- 11-year-olds on the topic of sound. This work falls fully within the constraints of the United Kingdom's National Curriculum;
- 12-year-olds' work in energy. This extra-curricular work was in response to a national environmental project established for schools by the World Wide Fund for Nature;
- 15–18-year-olds problem solving within a national award scheme (CREST) — some of which was inside and some outside of curricular time.

However, before I discuss the examples, I need first to comment quickly on problem solving.

Problems 'Open', Problems 'Closed'

Problem solving comes in a variety of formats. Broadly speaking, a closed problem is one where the solver is given both the aim or goal of the problem, most of the necessary information and the strategies for solving it. So, for example, an algebraic problem using F = MA, calculating the coefficient of friction between two moving objects, or the half-life of a radioactive isotope usually means that the student has most of the data and formulae provided. The goal is specified (for example, 'find F') and, while all the permissible moves may not be explicit in the problem statement, they are usually clearly defined elsewhere (probably in the preceding pages of the textbook, or in classroom notes).

With open-ended problems there are only very general principles suggested for finding the solution. Youngsters in the classroom must choose the best strategies from a wide range of possibilities to achieve their goal. These strategies are usually a combination of 'the design process' and 'the scientific method'. The main point of adopting this approach in schools is to require the learner to use a planned approach (their own or someone else's) to tackle a new problem based on their prior knowledge and learning and to produce a tangible outcome. It becomes their responsibility to delineate the problem, decide on what an appropriate solution might be, derive and test possible solutions and choose the point at which they think the problem has been solved. Their success is judged by the effectiveness of the solutions they develop.

However, the divide between open and closed problems is not sharp, but a continuum between types of problems.

The points in between the extremes of this continuum depict different styles so that, while students must still work within constraints from the closed end to the open end, they are increasingly given more freedom to operate. These kind of problems Munson (1988) has called 'curriculum-dedicated' — while they are more open, they arise from, and are embedded in 'normal' text-book physics activities in lessons. Some writers, like Garrett *et al.* (1990) have shown how it is possible to turn closed physics problems into more open ones. So, for example, rather than calculating the mechanical advantage and velocity ratio of a pulley system, it is possible to turn this around and present the system as a problem in lifting certain masses.

Qualitative Problems

The key issue here is that these kinds of problems are open to qualitative analysis, some forms of modelling and opportunities for clear experimentation. Some other examples might be:

- a plastic shower curtain is held by a rail across the entrance to the shower. When the shower is run hot, the bottom of the curtain blows inwards, towards the stream of water. Why? What factors affect the movement of the curtain? How can we test these?
- an upright freezer cabinet is opened to take something out. At the point of closing the door a gentle hiss is heard and the door is then very difficult to open again for a few seconds. After that point it opens easily. Why? What factors are involved with the difficult door? How can we test these?

These kinds of problems are based on everyday occurrences and some writers (for example, Bohren, 1987 and 1991) have produced a wealth of physics from the bubbles in a glass of beer and the light from a garage window — though lemonade and plane glass would work as well.

So what would the very 'open ended' extreme of the continuum look like? A number of projects in the United Kingdom have begun to explore this kind of work and I discuss one version here: a national project called CREST (Creativity in Science and Technology) for promoting problem solving in science for 9–19-year-olds.

The main intentions of this kind of work are to:

- explore pupils' ideas and provide tasks which encourage discussion and debate of 'how things work';
- develop problems so that different teams could work on different aspects of an overall contextual problem;
- build small group team-work so that individuals had roles within the team, which rotate during the course of the work;
- generate differentiated tasks and activities to cater for mixed ability classes;
- move between teacher-directed and pupil-directed activities and employ a variety of teaching and learning strategies;
- use a range of communicative techniques to focus upon both specific knowledge and higher level cognitive activities;
- promote conceptual change and the reconstruction of prior conceptions (Watts and West, 1992).

The youngsters work either as a whole class, in pre-determined 'science teams' or individually. They often work with a person from outside the school who is an 'expert' in the field. These experts are drawn from the local community — from industry and commence — and are a major strength of the way the scheme operates.

The model is one of cooperative learning and social collaboration. Following periods of work, individuals and groups share their solutions with the rest of the pupils. They work to an agreed set of criteria which form the basis of assessment — and discussion about progress. As they work, they become 'physicists for the day', tackling real problems. It is understood that sometimes it might not be possible to reach a fully-blown solution — but they will eventually reach feasible solutions for the problem.

Example 1: Sound at Spinfield School

Spinfield School is a co-educational school in modern buildings near a small town in rural England. The class in this study comprised thirty 11-year-olds (an almost equal gender split between boys and girls) and a highly competent class teacher. I was working with the school as part of a project to develop science in the curriculum (Watts, 1992) and was a 'guest-teacher' for one morning a week for seven weeks.

A 'real' problem associated with the school lies in the design of the school hall. This is a multi-purpose area which is used for school assemblies, for gymnastics and sports, and as a dining room at school lunchtimes. It is a concrete-glass-and-brick construction with a polished wooden floor and, not surprisingly, the acoustics in the room are very poor. During communal times the noise in the hall makes life a misery — even with no talk at all, the lunchtime clack of cutlery on crockery alone is deafening. It has all the acoustics of a swimming pool and teachers and pupils together dislike the area: raised or excited voices simply exacerbate conditions. The general problem, then: How to improve the acoustic properties of the school hall in a feasible and inexpensive way.

The concept of sound is part of the Science National Curriculum and, amongst other things, this requires that: pupils should study the way sound is produced and can be transmitted over long distances, how the ear works, the effects of loud sounds on the ear and the control of noise and sound levels in the environment. They should have opportunities to investigate the audible range and the relationships between loudness, amplitude, pitch and frequency. Pupils should investigate the effect on sound of the shape and materials of the built environment . . . (DES, 1991).

The Classroom

As indicated above, the main intentions of this work were to:

- explore pupils' ideas of sound and its properties and provide tasks which encourage discussion and debate of 'how sound works';
- develop sub-problems so that different teams could work on different aspects of the overall contextual problem;
- build small group team-work so that individuals had roles within the team, which rotated each lesson;
- generate differentiated tasks and activities to cater for a mixed ability class;
- move between teacher-directed and pupil-directed activities and employ a variety of teaching and learning strategies;
- use a range of communicative techniques to focus upon both specific knowledge and higher level cognitive activities;
- promote conceptual change about sound and enable the pupils to restructure some of their prior conceptions.

The youngsters worked both as a class and in their usual 'science teams' and the class teacher and I built in periods of team-work followed by groups sharing their solutions with the rest of the class. The overall problem of the acoustics of the school hall formed the raison-d'etre for the work: they all enjoyed being 'physicists for the day', tackling a real problem. They also understood the tentative nature of their solutions and were not disappointed at not 'fixing' the hall beyond doing a feasibility study.

In each session in the following seven weeks the science teams focused on sub-problems. The sub-problems consisted of:

- modelling the school hall both roughly and in terms of scale models;
- the design and efficacy of +string telephones;
- exploring sound insulation through designing industrial ear-muffs;
- studying the effects of loud noise on individual's abilities to concentrate and memorize;
- marking and measuring a +noise-trail through the school using a calibrated noise meter;
- constructing novel percussion and string instruments.

In the whole class mode they covered the content areas:

- a general appreciation of the nature of sound and vibration;
- particulate models of transmission and velocity of sound;

- the working of the human ear;
- reflection, echo, reverberation and absorption;
- amplitude and frequency;
- sound levels, noise, music and effects on people;
- sound in the built environment.

To do so they used 'slinkies', wave models, a signal generator, an oscilloscope and a variety of tape-recorders and sound-making devices to pursue the features of vibration and wave motion; studied the nature of vibrations through various materials, strings and examined musical instruments. They explored mechanisms for hearing, the biology of the ear and the acoustic range.

Pupils' Experiences

The science teams comprised five groups of six pupils each and these were their accustomed teams from past work. While each pupil had general responsibility for the overall task, each was also asked to take on a different role each session, which were:

- leader (arbiter of disputes, maintainer of group purpose)
- manager (site-manager, quantity surveyor, safety officer)
- go-getter 1 (fetch and carrier)
- go-getter 2 (fetch and carrier)
- writer (scribe, recorder of results)
- reporter (end-of-session class reporter).

Some parts of the lesson were teacher demonstrations, for example using the 'slinky' and other devices to explore wave motion, or using a class game to illustrate acuity of hearing. Sessions ended with a report by each group to the whole class, with a general discussion of what had been achieved, and where this now led. Groups were encouraged to communicate in a variety of ways and these included straight reporting, some creative writing, poems, poster displays, mime and role-play (Bentley and Watts, 1992).

The greatest difficulty faced in working with teams was deciding what help was needed, how much to provide, and when to provide it. 'Clue cards' of varying degrees of explicitness were written for each of the problems so that if the team was seriously off target, it was possible to offer one of these without further comment. After that, it was necessary to judge just enough intervention (beyond routine organization and management) to serve and challenge group thinking without providing solutions. 'Skill stations' were organized in parts of the room so that a particular practical skill (like using the glue gun, the oscilloscope, signal generator or the noise-meter) could be taught independently of team or activity.

Throughout, there was clear opportunity to explore pupils' ideas of sound and its properties and provide tasks which encouraged discussion and debate of 'how sound works'. Each team was asked to provide explanations and elaborations on:

- what sound is;
- whether or not, and how, it travels;

- how sounds are made;
- the role of solids, liquids and gases;
- molecular activity;
- reflection and absorption.

As other researchers have shown (for example, Scott and Asoko, 1990), the notion of velocity of sound was absent in the initial thinking of pupils. Similarly, absorption of sound proved to be very difficult for pupils to articulate — it relies upon some understanding of a molecular explanation of heat.

Example 2: Energy at Kingswood House School

This second study is drawn from a national event in the United Kingdom where schools have entered a competition for environmental problem solving sponsored by the World Wide Fund for Nature (a major environmental group). My involvement was first as a collaborator in the general school support for the event, as one of the judges of the competition, and then later as co-organizer of a writing workshop with teachers (Watts, West, Edwards and Champain, 1992). This work is an example of physics problem solving happening outside the main part of the daily curriculum. Kingswood is a small school for boys in rural England which has a very mixed ability intake. The teacher at Kingswood organized the group as an after-school club, a group of some eight or ten boys interested in science. One day in December they met to discuss what project they would enter for the competition. Initially there were very few ideas and very little 'brainstorming' was happening. It was a very cold day and one boy had brought a newspaper article about 'energy saving measures'. Like most in the United Kingdom, the school has a dining area serving hot meals to pupils at lunchtime. As the problem solvers looked around the school, they noticed the washing-up machine in the school kitchen was being flushed out and the hot water pouring into the drains caused a cloud of steam in the cold air. So the idea was born: they would explore aspects of the school to see what could be done to recycle energy. They would 'recapture' heat from the waste hot water from the kitchen and use it elsewhere.

They could not use the water from the machine outflow directly since it emerged under pressure and was not particularly clean (it had soap and food particles in it). They decided to use a 'reverse radiator' inside a large plastic tank as a heat exchanger. The water from the machines was directed into a large plastic dustbin as a heat 'reservoir', which then heated water in the 'reverse radiator' and which in turn was run into a copper coil inside a mini-greenhouse. They wanted to use this for the propagation of seeds: it was basically an insulated box on legs with copper tubing as the heating coil under the seed beds. There were several difficulties experienced along the way — for example, technical problems with plumbing and connecting plastic and metal tubes. At one point the machines 'back-siphoned', causing some flooding back inside over the kitchen floors (which the team had to clear). In this way the overall problem threw up a series of sub-problems, each of which needed to be delegated to one or two of the team. In some cases they needed outside help (from a plumber), sometimes tools, sometimes guidance in materials, sometimes help in the physics of making the converter more 'energy efficient'. They received assistance from other teachers in the school, from parents and from local people in the community — all of which the boys had to negotiate for themselves. They worked as a team with only some guidance and 'facilitation' from the teacher. Not only did they successfully complete the heat exchanger, they also tested its efficiency and effectiveness. As the teacher writes (Gollitt, 1992):

> The mini-greenhouse was tested by taking temperature readings inside and out, and also by growing mustard and cress. A 'control' box of seeds was put outside in the cold. Those died while the ones in the warmth thrived — even in cold January.

They worked hard and the problem overall was extremely motivating, The team *owned* the problem and wanted to see it through to successful completion. Throughout, the activity brought a 'torrent' of ideas from the youngsters who worked with it — there was no shortage of brainstorming once they were started.

The National Curriculum for this age range says (DES, 1991):

> Pupils should have the opportunity to compare and study a range of physical properties . . . including the thermal conductivity of materials, studied in everyday use. They should be introduced to the concept of energy by thermal processes and to the principle of energy conservation. They should have experience of a wide range of processes involving energy transfers in domestic contexts . . . they should be introduced to the idea of energy efficiency.

It is clear that this project met many of these requirements. The pupils wrote extensive project folders on their activities, drew on background information on global energy, local resources, alternative energy sources, appropriate technology in developing countries and so on. Even so, they were concerned as much with the physics of their 'reverse radiator' as they were with the wider issues, and clearly developed excellent answers to questions as to why they failed to get all the energy from the hot output from the machines. The teacher again:

> Following their success they were very pleased with themselves. They are now very, very enthusiastic about things which fire their imagination.

Example 3: CREST

The next set of examples come from the CREST award scheme and so it is necessary to say something about the project first. CREST is a scheme sponsored in part by the Department of Education and Science and industry, and supported jointly by the British Association for the Advancement of Science (BA) and the Standing Conference on Schools Science and Technology (SCSST). It is coordinated at a local level through Science and Technology Regional Organizations (SATROs) and by some local education authorities. It builds on a junior scheme called Young Investigators, also sponsored by the BA, and which is aimed at the 8–12 age-range. Both schemes complement normal school work and are non-competitive: youngsters gain recognition and reward for their work through the scheme as a national project, and receive a bronze, silver or gold award for their work. The scheme encourages (in its early phases) and requires (in its latter phases) students to identify and work on their own problems. These problems, though, are identified through an 'active partnership' between the students or school, and the local industrial/business community. Much of this depends on the flexibility of schools (and adaptability of teachers) in allowing students to pursue an investigative pathway — within limits.

CREST places considerable emphasis on 'problem identification', 'negotiation' and links between the student or schools and a range of outside agencies. Students are encouraged to develop their own strategies and, within the safety of the laboratory or workshop, are allowed to experience the successes (and failures) associated with project management. Negotiation at regular intervals over the criteria for success provides powerful insight for the students. They are not told what to do but are helped to achieve what they want in pursuit of their project's objectives. The quality of the students' experiences are monitored using a profile. This asks them to provide evidence which they consider demonstrates the CREST process criteria. The awards are criterion referenced: a series of 'You can . . .' statements on a record card are maintained by both students and teachers, pointing the way towards iterative problem-solving and successful project completion.

The three stages of the CREST award (bronze, silver and gold), then, are addressed by an accumulating set of criteria — a ladder of achievement in problem-solving process skills. Therefore, a major point of the project is that youngsters work on their own project. Individually or in groups, the project is one that they choose. Teachers may be influential at the point of choice – as may the CREST local organizer – but the emphasis is heavily weighted towards the youngster designing his or her own investigation. At the end, students are required to explain the development and outcomes of their projects to the outside agencies who have supported their work. The scheme works equally well for both teams and individual student effort. This is particularly so at the level of silver and gold where the scheme is looking for quite original (creative) work.

One of CREST's secondary aims is to promote closer working relationships between schools and engineering, industry and commerce in the 'outside world'. Some projects have involved the electronics industry where, for example, youngsters have designed a '7-day pet-cat feeding system' so that the cats' owners can leave the pet well fed while they are on holiday; or a system for allowing only the owner's cat through the cat flap.

Some CREST Projects

While these projects are referred to as individuals' problems, they are commonly the product of group work in their earlier stages.

- 'Clearway'
  A baby, born five minutes earlier at a hospital in Nottingham, receives the benefit of a disposable 'Clearway' mucus extractor. The 'Clearway' extractor clears the baby's airways and minimizes soreness, risk of infection and replaces oral suction techniques and the possibility of cross-infection from nurse to child and back again. The device is a simple

single-hand-operated pump-action container to clear the nasal and bronchial passages. It is made of plastic and uses a concertina bellows and a butterfly valve. While it was developed as part of Angus's examination work for GCSE, when he was 15, it also won a gold award through CREST. The device was developed through his work experience at the hospital and later with some advice from a United Kingdom manufacturing and distribution company. It is now being adapted for veterinary work for use on sheep, cats and dogs, clearing the oropharyngeal passages at birth.

- Bicycle digital speedometer
  Andrew won a gold award for the design of a digital bicycle speedometer. This is a non-friction device using a light gate which reads a slotted-wheel each time the bike wheel rotates. The prototype was quite large and Andrew saw his task being to miniaturize the system so as to reduce bulk and weight for cyclists. The work was helped by members of the Physics Department of the local University in collaboration with his school in Belfast, Northern Ireland. His device is now of interest to a local electronics manufacturer.
- Helicopter control
  This work was conducted in cooperation with a helicopter design company at Costock in the English Midlands. Because it is impossible to carry out research on full-sized helicopters, Russell built his own radio-controlled model. The work focused on moving forward the centre of gravity of the rotor blades. A strip of lead was put inside the leading edge of each blade which made them heavier and increased the inertia of the rotor head. This added to the helicopter's stability in hover mode and made it more difficult to blow off-course in gusting wind. From this, other modifications were made, including reducing the mechanical 'slop' of the linkages and reducing the pitch of the rotor blades to make them increase their speed and control response in flight.
- An automatic referee
  Simon's project grew out of his love for windsurf racing. During a race the point is to manoeuvre round a series of buoys — without touching them. Touching the buoy is penalized by having to turn and round the buoy again cleanly. Normally the referees have to watch through binoculars to decide if the buoy was touched but Simon's device is automatic. The electronics are built into the buoy itself and can distinguish between wave action and a distinct collision. Not only does the device set off an alarm on the buoy, it also sends a radio signal to the referees on shore and can activate a video camera to capture the event on tape. The work was developed alongside a local company who are now interested in developing the product.
- Rainbow arcs
  A less technologically orientated project has been Annette's study of rainbow formation. She has designed a novel approach through balancing drops of liquid on the end of a light-fibre, allowing investigation of liquids of different refractive index. She then developed a 3D geometrical computer model from the results, considering nth-order diffraction patterns for raindrops. The work was conducted in collaboration with members of the Physics Department at her local university.

- Rapid evolution
  Anne-Marie is the winner of two CREST gold awards. Her major interest is in designing computer programmes. Although the topic here is a biological one, the computerization of systems is a common thread across sciences. Anne-Marie set herself the task of simulating evolution by natural selection — in this case the sexual and asexual reproduction of worms, the exploration of the replication and mutation of a limited genetic code. She is based at a school in Northern Ireland, achieving her second gold award for a sophistication of the first — an approach to a 'neutral computer' through pattern recognition.

It may help here to note just a few of the other problems that have successfully been tackled through CREST:

- a weather satellite receiver;
- a remote controlled satellite dish;
- an electronic +adder for very young children;
- a demonstration heat-pump unit;
- a system for automatically watering house plants in the home;
- a pulsating magneto-hydro-dynamic generator;
- a safety kettle for a blind person;
- a battery powered vehicle to seat 4/5 people;
- an infra-red proximity detector for lift doors;
- a digital flow meter to measure water flow in a river.

There are many, many more — both in physics and in the other aspects of science.

Forms of School Organization

Before leaving these examples fully, it is worth noting how task-orientated problem solving can be organized in schools. It can be seen from these kinds of examples that there are a number of ways for problem solving of this sort to take place (see also in Watts, 1991). In general, the forms are:

- Activity days — in some schools a group of science and technology teachers will suspend the curriculum for the day and instigate a 'problem-solving day' for groups of pupils. The problems are negotiated in advance so that resources and materials can be planned;
- National competitions — some schools have been involved in national competitions (for example the Esso 'Egg-race' challenge, or the World Wide Fund for Nature Environmental Enterprise Award) and have used this as the basis for forming groups for problem solving. The groups have been whole-class groups or 'after-school' clubs;
- Curriculum problem-solving — some curriculum materials (for example SATIS and Process Science) encourage teachers to develop problem solving as part of the everyday curriculum in class. While this is often related to the physics curriculum in hand, it also generates questions and research projects which pupils want to develop;

- Activity weeks — it is not unusual for schools to suspend the curriculum for all pupils at the end of the summer term when most assessment procedures are complete and allow the youngsters to choose from a range of activities. In this instance scientists and technologists join to provide pupils with a range of opportunities to tackle problem-solving;
- Summer schools — in some areas there has been considerable collaboration between schools and universities to develop residential summer schools for pupils to tackle problem solving. Here there are often staff and graduate students on hand to work with the groups, using the university's facilities.

Constructivism and Cognitive Change

So, what can be said about the virtues of this way of working and how does it fit generally with constructivism and conceptual change. It is possible to be purist about constructivism as I explore in a moment, though it is also known to be a fairly broad church which encompasses writers of Ausubelian, early-Piagetian, Kellyian and Vygotskyian persuasion (to note just a few). Moreover, there is now a growing number of case studies which document the use of constructivist approaches in science classrooms, as described elsewhere in this volume.

In my view, constructivist learning is always an interpretative process involving individual's constructions of meaning relating to specific occurrences and phenomena. New constructions are built through their relation to prior knowledge — and the pedagogic challenge for teachers is to focus on students' learning-with-understanding, rather than the more common (and straightforward) emphasis on 'covering content'. To learn science from a constructivist philosophy implies direct experience with science as a process of knowledge generation in which prior knowledge is elaborated and changed on the basis of fresh meanings negotiated with peers and teacher.

In the work described above, it has been important to promote learning of this kind relating to a variety of concepts in science. In the United Kingdom's current post-National Curriculum era, the 'coverage' of the content of the statutory 'Programmes of Study' is of immediate importance. The challenge undertaken has been to meet these legal requirements, while at the same time fostering a constructivist approach through active problem solving strategies. Before taking this further, I want to dwell on constructivism a little more.

Constructivist Principles

In this section I want to explore two main points:

- though there is a considerable amount of research within constructivism, there is little by way of strict comparison between 'non-constructivist' and constructivist approaches;
- as Layton notes, there is little that explores in detail 're-constructivism' comprising the transfer and application of understanding in *real* situations.

Let me take the first point first: it is possible (Watts and Bentley, 1991) to make a distinction between 'strong' and 'weak' constructivism, as follows. In my view, 'strong' constructivism centres upon:

- Cognitive construction — the heart of constructivism is the view that cognition is the result of proactive mental construction. Conceptualization, it is said, arises through the interaction between previously accumulated knowledge and current data, and — as Simpson (1990) points out — it is 'constantly at work, testing and ascribing meaning' to new information in terms of individuals' prior conceptions of phenomena in the world. This is an emphasis upon active, anticipatory, whole-bodied, 'form-giving' (Mahoney, 1988) cognition;
- Constructive processes — these are such processes as construction, deconstruction and reconstruction (Bruner and Kenney, 1965). They imply that cognition has structure and organization. This is commonly described (Gilbert and Swift, 1985) in terms of core or deep structures which are predominately tacit, less amenable and accessible to change, and the more explicit, peripheral or surface structures of ongoing activity. The emphasis here on processes is upon the to-and-fro between deep and surface structures as they slowly adapt through interactions in everyday life;
- Oppositionality — as we construct and qualify meanings, we do so against a backdrop of comparison or contrast with other meanings. Constructivism suggests pluralism and a relationship *between* ideas rather than simply the development of ideas de novo: concept development takes pride of place over concept formation;
- Critical realism — constructivism views knowledge as transitory and provisional. Knowledge of the world is constructed on the basis of the constraining influences of the nature of phenomena, personal context, language, predisposition, etc., and judged by such criteria as utility, plausibility and fruitfulness (Strike and Posner, 1985);
- Self-determination — constructivism implies a metacognitive position which tends to take two forms: for teachers it is termed 'reflection', for students it is 'learning about learning'. While there is debate on the one hand about the nature of reflection (between, say, reflection and contemplation: Buchanan, 1990) and the commitment of pupils to metalearning on the other (White, 1988), there is a clear strand through constructivism that the person at the centre of the enquiry is not just an 'active meaning maker' but knows s(he) is too. That is, constructivism sees human actions, including learning, to be purposive, consciously aimed towards some end;
- Collegiality — the 'being in the world' of constructivism implies a social context where ideas and conceptions are communicated, shared, tested, negotiated, and reported. There is also a sense in which constructivism implies caring — caring for ideas, personal theories, self-image, human development, professional esteem, people — it is not a take-it-or-leave-it epistemology.

So *good* constructivist practice in classrooms ought to be different to good classroom practice generally in distinct ways. The emphasis will be on active learners constructing and reconstructing their own ideas, taking responsibility for their learning in ways they know they can do, being self-determining within a caring group, negotiating with others towards purposive ends.

While there have been a number of reports on classroom work within a constructivist framework, they have focused on some but not all of these aspects. Moreover, they stop at the point when learners begin reaching understanding, rather than the points of transfer and application of ideas. To return to Layton's diagram (figure 4.1) for a moment, much of the research and development in this field has concentrated on the first three sections: learners' everyday knowledge, the processes of science education and knowledge for understanding. There is nothing wrong with this except the feeling that it stops too soon and does not take in the next sections. Elsewhere, I have reported some explorations of pupils' solutions for individually negotiated, open-ended, talk-orientated, qualitative problems in science (Watts, 1991). My underlying thrust in this work is to encapsulate as fully as possible a cluster of constructivist strategies. In this approach, the classroom work skirts the borders between investigative science and technological capability. A firm evaluation of this work in schools is still in progress, but the examples I have cited do help to resolve some of the questions I have raised.

Learning in Action

There is little doubt to me that the kinds of pupil experiences described above resulted in quite considerable amounts of learning. The solutions that pupils arrive at — whether for the insulation of a room, the heating of a seed bed, or the design of a helicopter blade — have not happened accidentally. Some of this learning has been charted in formal and semi-formal ways (not least of all for National Curriculum assessment) and — in the case of the CREST problems — this has been recorded and acknowledged in very clear and public ways.

In problem solving, pupils have implicit theories-in-action about how things work. They are not always asked to verbalize these but — in order to reach a solution — they must 'realize' and give expression to their theories in some deliberate way. The most powerful impact of problem solving is in the motivation it gives through tackling real problems, of taking ownership for the issues so that, in the full flush of activity, youngsters are barely conscious of the points when they articulate their 'alternative conceptions'. So, for instance, at Spinfield the ideas about sound, air, molecules and energy tumbled out of the pupils as they tried to debate amongst themselves just how they could make the insulation more effective. Within the throes of the classroom work it was possible to predict and plan for some of these conceptions, while others needed to be tested and dealt with as they arose. Some, also, had to be shelved and could not be touched at all since to do that would have taken the steam and direction away from the main task in hand.

Similarly, the work on energy transfer at Kingswood raised all the many alternative conceptions known to exist in that area. Was energy really travelling out of the hot water, through the tubes and then — eventually — directly into the seeds? The discussions allowed the teacher to explore the many fundamental aspects raised through the focus of the task in hand. The classroom or school problem solving activity, then, becomes an arena where this explication, exploration and evaluation can all happen with a distinct and relevant purpose — so that the task at hand is not some disembodied learning within science, but that learning occurs as a real means to a real end.

Constructive Processes

So, do these problem solving activities make for 'weak' or 'strong' constructivism? Needless to say, not all the teachers associated with the examples above would admit to relating to either — even if the question had been asked of them. There is no sense in which they have been acting within any consortium of principled pedagogies, constructivist or otherwise. However, the question can be asked of problem solving per se and here there is the clear possibility that the strong form can (and has been seen to) emerge. Let me revisit just three of my principles here — construction, reconstruction and collegiality — and leave others (like realism and metacognition) to other authors in the volume.

First, conceptual construction. As pupils begin the process of sorting the activities within a particular problem, it is possible to see them explore the 'conceptual space' of the problem itself — to reach some sense of the 'size of the task': what needs to be done, how it might be achieved, what the overall requirements are, what it is all about. They then need to 'import' into this space their understandings of how things work along with their theories, associated ideas, playful notions, moments of whimsy. They trade on both domain specific and domain general knowledge, constraints, principles, causation, beliefs, probabilities and perspectives. The search for a solution (*their* solution) becomes a driving force behind their need to construct, reconstruct and fine tune their theories. So, for instance, within the specific domain of how sound is mechanically transferred, youngsters need to add their understandings of the effect of this on the ear, on their powers of concentration, whether they can do homework while listening to music, watching television, etc., the effects of an ear infection on hearing, and so on.

There is an unanswered question here of how best to map this space and chart the many theories, conceptions and associations which are used within it. This is indeed an intriguing project since the possibilities are that such a mapping would be much more complex than, say, simply explicating pupils' conceptions and theories around a single concept like 'force', 'matter' or 'living'.

Second, conceptual processes are clearly in evidence too. Let me ride the issue of reconstruction for a moment. In Layton's diagram, reconstruction is the aftermath of construction. In terms of problem solving, this is the stage where the conceptual frameworks have been drawn into the conceptual space and have been made explicit and explored for their use, importance and implications. That is, if particular theories about sound and molecular activity have been described and examined, these have then to be re-mixed with other knowledge (about materials, nuisance levels, social habits, etc.) but then they must be translated and reworked into a fitting solution for the problem. The end is not reached simply by having a new (more scientific) conception, but rather by putting that to use in a very tangible way. Many of the conversations with pupils as they work give grounds for appreciating the constant to-and-fro between their surface activity and the deeper reservoir of understandings upon which they draw as they work.

This opens to another unanswered question about the nature of intuitive theories and their explication: just how implicit can a theory be and still be a theory? There are many occasions during problem solving when youngsters do not fully explicate their theories — either because they lack the time, opportunity, or because they have just too many crowding into the 'space' to separate and differentiate them all. However, there are clear points during their work when pupils will express a preference for particular types of mechanisms relating to phenomena with which they can have little acquaintance — and to that extent can be said to have theories about them. For example, they can suggest how the surface properties of a material or structure relate to the reflection and transmission of sound.

Third, collegiality seems to place emphasis on social construction of knowledge. There is no doubt that the construction and reconstruction of knowledge takes place as an interplay between the directions, motives and purposes of the individual within the larger directions, motives and purposes of the immediate group or social situation.

However, the inclusion of 'collegiality' here is not to attach undue weight to either the individual or the social construction of knowledge — both deserve separate and concerted exploration. While studies of 'alternative frameworks' in the past may have focused too tightly on the individual (as Solomon, 1992 suggests), individuals are still an important area of work. Collegiality does, however, acknowledge social constructivism and a strong classroom version would provide all the opportunities possible for pupils to enhance conceptual change through collaborative interaction between peers and teachers. Within the types of problem solving described here it should be clear that this has been a constant thread throughout all the activities. The question (again unanswered) that arises here is to what extent conceptual frameworks are mediated through interchange. Is there a spectrum so that, at one extreme, some are so malleable that they undergo revision within each (focused) conversation while, at the other, some are so durable that they escape change regardless of the interaction? This is really a question about the 'glue' that holds alternative conceptions together with any sense of coherence. In my view, task-orientated problem solving provides an excellent test ground for exploring the nature of this glue: seldom do students have to work through the consequences of their thinking to the final delivery of a solution to a complex problem so that it 'all comes together in the end'.

Summary

One of the challenges for physics teachers in the United Kingdom is to meet the legal requirements of the National Curriculum while at the same time making science relevant and creative. In this chapter I have described some of the work in progress which focuses on learning-with-understanding. In my examples the emphasis is not just on 'covering content' in a school syllabus but of motivating students to apply their knowledge and understanding.

Our aims and ambitions for science education are fairly wide and all embracing and no single approach to teaching and learning can possibly do them all justice. My comments in this chapter have been directed towards one particular pedagogic approach: the use of open-ended, task-orientated problem solving.

The central virtues of problem-solving rest on its potential to both facilitate the construction of knowledge and its transfer of learning from one context to another, and to encourage the transfer of responsibility for learning from teacher to learner. Problem solving of this kind is less constrained by didactic teaching methods. There is no suggestion that this should be the sole way of working in laboratories or classrooms. I would argue, though, that without some inclusion of work of this sort we will never come close to the application of science to students' life worlds.

The examples I described clearly concern the application of scientific facts, skills, concepts and attitudes — and there is evidence that these change in nature and develop in quality during the course of the activities. Students' cognitions change as they work through the 'conceptual space' surrounding the problem and as they realize their solutions in 'hardware' outcomes. The examples begin with a greater degree of specificity than they end and I have spared little space for a detailed discussion of Layton's (1991) process of deconstruction. The reconstruction of the science involved in each case is manifest in the reports, discussions, negotiations and technical outcomes of the groups and individuals as they work.

Broadly, I think this way of working does approach some of the central principles of constructivism. In general, it can be seen to:

- provide opportunities to explore and elaborate pupils' naive and developing understandings of science;
- promote active learning, and 'actionable' learning — where pupils must put their understandings into practice and use their knowledge;
- engender shared teamwork and collaborative group activity within an overall social context;
- work through the use of open-ended investigations, where there are few 'right answers' and approaches to any solution — where pluralism rules;
- make science relevant, enjoyable, fruitful, plausible and highly motivating.

The examples were also intended to show that:

- teaching science through practical problem-solving is both possible and profitable. Topics like sound and energy can be taught through these methods. The teachers involved are not super-human;
- over time it is possible to access pupils' knowledge and skills in science through this form of work. It is not as easy to do as with a written test, a 'calculations assignment' or an examination. However, in this case their knowledge is knowing-in-action and a valuable part of schooling. The progress of each group was measured in terms of their success in solving the problems set, in articulating their goals and accomplishments, and in their appropriate use of the language and elaboration of the science involved.

References

Association for Science Education. (1981). *Education Through Science.* A policy statement. Hatfield: Association for Science Education.

Bentley, D. and Watts, D.M. (1992). *Communicating in School Science. Groups, Tasks and Problem Solving, 5–16*. London: Falmer Press.

Bohren, C.F. (1987). *Clouds in a Glass of Beer. Simple Experiments in Atmospheric Physics*. New York: John Wiley and Sons.

Bohren, C.F. (1991). *What Light Through Yonder Window Breaks*? New York: John Wiley and Sons.

Bruner, J. and Kenney, H.J. (1965). Representations and mathematics learning. In Morrisett, I. and Visonhaler, P. (Eds.). *Mathematical Learning. Monographs of the Society for Research in Child Development*. Serial 99, *30*(1), pp. 50–9.

Buchanan, M. (1990). How practical is contemplation in teaching. In Day, C. Pope, M. and Denicolo, P. (Eds.). *Insight into Teachers' Thinking and Practice*. London: Falmer Press.

Carmichael, P., Driver, R., Holding, B., Phillips, I., Twigger, D. and Watts, D.M. (1990). *Research on Students' Conceptions in Science: A Bibliography*. Leeds: Children's Learning in Science Project, CSSME, University of Leeds.

Department of Education and Science (1991). *Science 5 to 16 (1991)*. London: HMSO.

Garrett, R.M., Slatterly, D., Gil Perez, D. and Martinez-Torregrosa, J. (1990). Turning exercises into problems: an experimental study with teachers in training. *International Journal of Science Education*, *12*, pp. 76–83.

Gilbert, J.K. and Swift, D.J. (1985). Towards a Lakatosian analysis of the Piagetian and alternative conceptions research programmes. *Science Education*, *69*, pp. 681–96.

Gollitt, V. (1992). Case Study 3: Energy at Kingswood House school. In Champain, P. Edwards, P. Watts, D.M. and West, A. (Eds.). *Making the Difference: Environmental Problem Solving Through School Science and Technology*. Godalming: World Wide Fund for Nature, United Kingdom.

Layton, D. (1991). Science education and praxis: the relationship of school science to practical action. *Studies in Science Education*, *19*, pp. 43–79.

Mahoney, M.J. (1988). Constructivist approaches in educational research. *Review of Educational Research*, *47*, pp. 651–93.

Munson, P. (1988). Some thoughts on problem solving. In Heaney, J. and Watts, D.M. (Eds.). *Problem Solving: Ideas and Approaches from the Secondary Science Curriculum Review*. Harlow: Longmans for the Schools Curriculum Development Committee.

Pfundt, H. and Duit, R. (1991). *Bibliography: Students' alternative frameworks and Science Education*. IPN Reports in Brief. Germany: University of Kiel.

Simpson, M. (1990). Why criterion-referenced assessment is unlikely to improve learning. *The Curriculum Journal*, *1*, pp. 171–83.

Scott, P.H. and Asoko, H.M. (September 1990). *A Study of Students' Understanding of Sound, 5–16, As An Example of Action Research*. Paper presented to the Conference of the British Educational Research Association, London.

Solomon, J. (1992). *Getting to Know About ENERGY in School and Society*. London: Falmer Press.

Strike, K.A. and Posner, G.J. (1985). A conceptual change view of learning and understanding. In West L.H.T. and Pines, A.L. (Eds.). *Cognitive Structure and Conceptual Change*. London: Academic Press.

Watts, D.M. (1991). *The Science of Problem Solving*. London: Cassells.

Watts, D.M. (1992). *Zones of Sound: Extending Problem-solving in School Science*. Monograph. London: Roehampton Institute.

Watts, D.M. and Bentley, D. (1991). Constructivism in the curriculum. Can we close the gap between the strong theoretical version and the weak version of theory in action? *The Curriculum Journal*, *2*(2), pp. 171–82.

Watts, D.M. and West, A. (1992). Progress through problems, not recipes for disaster. *School Science Review*, *73*(265), pp. 57–64.

WATTS, D.M., WEST, A., Edwards, P. and Champain, P. (1992). *Making the Difference: Environmental Problem Solving Through School Science and Technology.* Godalming: World Wide Fund for Nature U.K..

WEST, A. (1991). Science and technology, In WATTS, D.M. (Ed.). (1991). *Science in the National Curriculum.* London: Cassells.

WHITE, R.T. (1988). *Learning Science.* Oxford, Basil Blackwell.

Chapter 5

Structures, Force and Stability. Design a Playground (Year 6)

Cliff Malcolm

I offer a story of a unit of work in which year 6 children design a playground for children in years 1–3 and submit their proposal to the school council for possible implementation. The story is not an actual case, but a composite of many cases of teachers in Australian primary schools who are experimenting with approaches like this. Thus it draws directly on research and teaching-learning in actual cases.

The story could indeed be true — witness the case examples recounted by Brass *et al.* (this volume). Playground design is an activity tried by many teachers, and is recommended in the New South Wales *Science and Technology K-6* (1992). Schools around Australia have extensive freedom in curriculum design; elected School Councils and parent committees have established responsibilities in school development (including curriculum planning); students, especially senior students (in both primary and secondary schools), are involved in school leadership and make real contributions.

By telling what the teacher and children did, what the setting was, what happened, I have given the unit a sense of time and place, and demonstrate the central role of teachers in curriculum and the complexity of their roles in the classroom. As well, the story lays open the complexity of the task of curriculum design. Planning curriculum is more than asking 'What is the best way to teach X?'. There are so many X's and they have to be all considered at once and all in the context of school management: theoretical concepts, practical skills, cognitive skills, metacognitive skills, social skills, self-management; insights into the natures of science, learning, other people, social structures, self, nature, God. Curriculum design has to enable these various outcomes to interact, reinforce, contrast and confront each other, like characters in a play. But as with a play, there must be a plot, a sense of going somewhere, a meaning, and confidence (from students and teacher) that the journey is worthwhile (see Wittrock, chapter 3 in this volume).

In *Design a Playground*, the plot comes from the design question. The problem is a quality of life issue, seen as such by the children and the school council. Solving it draws authentically on the children's experience and follows through for them to personal action. Science arises in the context of the problem: theories about structure, force and stability; processes of designing and planning, collecting and analyzing information, drawing conclusions, providing evidence; work skills of management, cooperation and working alone. Questions of what leads what — whether process leads product, technology leads science, personal development leads concepts or some other combination — do not arise; the various facets are integrated and interactive.

I intend the story to be useful to teachers and education students mainly as a discussion starter, to debate the quality of the example and raise alternatives; question the implicit view of science and the balance of science, technology, society and personal development; analyze the roles of the teacher and the implications for teacher education. A complex story such as this can be held under many different lenses — what the teacher does, what students do, how does the teacher monitor classroom processes, assessment, intellectual climate, facilities and timetabling, educational priorities and responsibilities for planning.

I invite you to question the value of stories and cases like this, and investigate ways of presenting and using them to effect. One difficulty, for example, is that, when ideas are presented in realistic context, the context can interfere with the message — for example, a reader who finds designing a playground trite in comparison with, say, an issue like water quality, might have difficulty seeing past the context to the comment on curriculum and teaching. At least we need other stories in other contexts, such as are offered for example, by Driver (chapter 15 in this volume), Brass *et al.* (this volume), and Malcolm (1987 and 1992a).

Description of the Unit

The story of the unit is told below in the left-hand column. Theoretical and research underpinnings are on the right. The major research bases are from the interrelated set of:

- constructivist learning theories (including metacognition);
- linking theory to practice, social purpose and individual development (*STS-P: Science, Technology Society and Personal Development*);
- inclusivity (*Science for All*);
- development of broad skills, such as working cooperatively and problem solving;
- coherence and progression in learning outcomes.

These principles are policy for science education in Australia. Traditionally, primary science gives strong attention to process skills and students' personal development. The unit below builds on and extends this tradition.

Getting Started

In the wake of Monday's school council meeting, Joseph explains to his year 6 class that the council wants to set up in the space across from the Art room a playground for years K-3. A subcommittee will consider possibilities,

The problem solving form is one basic structure for an STS-P approach (Malcolm, 1987; Fensham, 1988). See also Watts, (chapter 4 in this volume). The problem serves as an entree to an integrated treatment of science, technology,

and wants to involve students in the discussion.

society and personal development (STSP). The 'society' for these children consists of the school community, and their social concern is the quality of life that the community can provide for its members.

The problem to be solved gives interest, coherence and direction to the unit; it defines the 'plot'.

'What equipment do you think we should have?' Joseph asks. The students think back to their favourite playgrounds. They create a list on the blackboard, trading memories as they go. Joseph asks each student to choose a piece of equipment from the list and write a short argument, 'Why I think the playground should have a . . . .'

Joseph's opening question is divergent and invites students to reflect on their experiences. It has the capacity to engage and involve all students. It sets the problem in a way which values students' opinions, creates expectations, and builds a working team (Baird and Mitchell, 1986; Johnson and Johnson, 1990).

Clarifying the Problem

'If you were the engineer asked to design this playground, what would you need to consider as part of your planning?' Joseph probes and questions the children to help them think through the problem. Their list includes fun, size, graffiti, costs, strength, won't tip over, shape, traffic jams, safety, drainage, materials, painting.

As well as involving students and building a sense of anticipation, Joseph is building an intellectual climate of thoroughness and expectation (NCTM, 1991). He reinforces this by having the children write down their ideas. The writing time allows him to move around and monitor children's progress.

Joseph works back through the list, exploring what the terms mean and why the considerations are important. What makes a playground 'fun'? Has anyone had an accident at a playground? How is drainage a problem?

The children in pairs draw up tables of 'Things to consider' using headings *fun*, *structure*, *safety*, *appearance* and *cost*. They classify the ideas on the board and add others.

Joseph moves around, listening and talking. The children are used to brainstorming and classifying, but many are still bothered when things fit in more than one category. Karen is especially bothered that activities which seem a lot of fun tend to be dangerous. Joseph chooses not to follow up the idea at this stage.

Negotiating a Plan

Joseph asks the children whether they would like to submit a plan to the school council. He explains that he has talked with the Principal and the Council President, and they are supportive. He outlines what he thinks will be involved, using the lists on the board, and talking about the ways that engineers work. The children are enthusiastic.

The children are part of the planning. They see the scope of the problem and themselves in relation to it. Joseph encourages and supports them. He knows that he is leading them, and is trying to gauge the level of their commitment. Part of this is to help the children see clearly the work that is involved and how it might be done. He links the planning to the work of a 'real engineer' to help the students build a picture of science.

Joseph asks them to reexamine their 'Things to consider' lists and think of ways they can collect the information they need. Under 'fun' they want to visit some local playgrounds, watch children at play, interview them. Under 'safety', they focus not on equipment breakage but collisions and falls. To explore 'construction' they suggest looking at the ways equipment is made, building models and testing them. They are unsure about how to investigate 'costs'. Perhaps someone could go to a shop somewhere.

Joseph promises to put their ideas together in a plan they can consider next Monday.

Discussing the Plan

Joseph works on the plan first by himself, starting from the children's ideas, drawing concept maps and flow charts, thinking about resources and management. He talks to the Principal and Council President, and at length with two other teachers, the curriculum coordinator, and the year 1 teacher. He

The plan (figure 5.1) brings together aims, content, context, teaching methods and assessment. Joseph holds the focus on the project goal. It provides the main plot. Activities, content and learning goals all relate to that story (Malcolm, 1992b; Watts, chapter 4 in this volume).

knows the work will take a few weeks and that the class will need time in two-three hour blocks. He schedules two half-days per week for six weeks. The father of one of the children in the class is an engineer, and Joseph talks with him about some of the ideas, especially ideas of stability and deformation.

His resulting plan is given in figure 5.1. He writes it in terms of two sets of goals — activity goals (what the chidren will produce) and learning goals (what they will learn). For the children to learn through cooperative work, Joseph knows that good management is essential, so he writes management plans as he goes. The next step is to talk the plan through with the children.

Joseph knows his plan is incomplete. He has left out costs. He is unsure of how far to go with the investigations of strength and stability. Should he stop once he has the patterns or rules? Should he try for theories that explain the rules, or theories that explain the theories? He doesn't know how to protect the chidren from the fact that their design might not be implemented by the Council. He is not sure of how much time will be required.

He hasn't planned assessment fully, but he knows the basis is there — students will do things he can watch and produce things he can assess. Children will often work without him, and evaluate their progress and learning as natural parts of their work. He has given strong emphasis to the production of posters and models, anticipating the presentation that the group will make to the Council and parents.

He talks the plan through with the children, presenting it optimistically, setting it into the context that 'this is how engineers would do it'.

Joseph is following an approach to planning sometimes called 'work required learning' or 'goal based learning' (Ministry of Education, 1988; VCAB, 1992). The work to be done is specified through activity goals defined broadly enough that all children can come in at their own level and flexibly enough that individual learning styles and interests can be built in.

To control the length of the project, Joseph uses parallel tasks. Jigsaw strategies and whole class reporting systems keep the ownership with the whole group as well as requiring children to teach one another, and put their ideas up for scrutiny by their peers (Dalton, 1985).

The students are to be involved in the planning, and anyway, the task is divergent, so Joseph's primary concern is to anticipate the management requirements rather than the details of questions and answers. Johnson and Johnson (1990) call this 'conceptual application' (operating from general principles) as against direct application (operating from tightly defined plans).

Joseph plans the assessment as part of planning the work. He builds in opportunities to observe the children at work, and requires productions that can be appraised (Lovitt and Clarke, 1988).

Figure 5.1: Goals and plans for the unit

Activity goals	**Learning goals**
1 Analyze playgrounds • develop an observation schedule and eventually a poster noting construction, design and layout of existing playgrounds • develop an observation schedule and eventually a poster showing what equipment children spend their time with and what they do • develop an interview schedule and eventually a poster summarizing children's opinions	• data and analysis are important in problem solving • planning for, collecting and presenting data • designing and making a poster
Management Each student belongs to two groups, a home group and an expert group. There are three expert groups (one for Structures, another for Watching and a third for Interviewing) each with three members. Home groups also have three members, one from each of Structures, Watching and Interviewing. In the home groups, each member teaches/informs the others about progress in his/her expert group In the expert groups, three roles are assigned: — records manager (to monitor the note taking) — task manager (to monitor planning and achievement) — team manager (to monitor group processes).	• working together • communicating ideas • personal management
2 Decide what pieces to build • produce overall summaries of the analyses in 1 • prepare written reports on safety versus fun • write a whole-class plan saying what pieces will be built, what scale will be used, and how the work will be organized	• reporting to a large group • understanding cost-benefit ideas
Management • seek agreement on the best 6–8 pieces/components to build. (Postpone discussion of a 'grand plan'). • discuss the merits of integrated designs versus separate pieces. • set up pairs to build separate components/pieces. Plan towards three total models—i.e., it's alright to have three or four slides, but we also need a range of pieces. • *After* the model pieces are made, look to ways they can be combined onto the space and hence produce two or three 'grand plans' (see below)	• decision-making in/for a large group • team skills—cooperation and compromise

Figure 5.1: (Cont.)

Goals and Unit Plans (cont.)

Activity goals	Learning goals
3 Preliminary design • invent three possible designs for your piece, and show them in drawings. For each design write down good and bad points • complete experiments to investigate principles of stability and strength and demonstrate your findings. • make summaries of the rules of stability and design for Shapes, Beams and Balance	• creating and evaluating options • designing and conducting experiments • understanding stability and strength in relation to balance, shape, struts, beam shape, spans/cantilevers
Management Children do their designs individually. We 'step sideways' into investigations: use the jigsaw strategy as before (same home groups and work groups), with the expert groups working on Shapes, Beams and Balance.	• working alone • working together
4 Construct the model • develop and draw your favoured plan, showing its measurements. Write explanations of its strength, safety and play value, like an advertisement. • construct your model (from drinking straws, pipe cleaners, satay sticks, card, tape, plasticine . . .) • write advice on how the 'real' piece should be made (talking about materials, beam shapes, etc.)	• drawing to scale • construction skills • limitations of modelling and scaling
Management Children work in pairs. Three equipment tables needed.Give the students a written *Design Brief* for this step.	• managing equipment and construction
5 Put the pieces together • produce alternative layouts for the playground, assembled from the models. • write a report suggesting likely traffic flow, and commenting on safety and fun value.	• spatial relationships • design, evaluation
Management • Aim for three alternatives, starting with pieces which have been made by more than one group • Mark out areas on the floor that simulate the play area and place the models	

Figure 5.1: (cont.)

6 Prepare the submissions • produce three submissions to council. • note in the submission that costs and choice of materials have not been studied, but will be important in the council's decision. **Management** • use the three models accompanied by display boards showing a collection of the posters, charts, etc that were produced during the project and tell the story of the project. • be aware that Council may choose not to implement any of our models; justifiable from costs and materials • set up the displays for council, parents and other children to see	• organizing and communicating information

Analyzing Playgrounds

Joseph knows that taking the children to local playgrounds will be a major exercise, requiring a supply of small children and adults to help with supervision. He will ask Maria and her year 1 class. He can arrange parent help from year 1 or year 6. There are two playgrounds in easy walking distance but the next closest is two kilometres away. Meanwhile his class has to prepare for the visits.

He explains what has to be done and how it will be organized. Children choose to be part of the Structures group, Watching group or Interviewing group, and with Joseph's guidance the groups are adjusted so that they are equal (nine members each). Within these groups, Joseph nominates the groups of three and the home groups.

He assigns at random within expert groups the roles of Records Manager, Task Manager and Team Manager and talks with the class about the roles. They

Work required learning and cooperative groups place heavy demands on work management and people management in comparison with traditional teaching.

Children need to understand the goals of the work, management structures and processes (Lotan and Benton, 1990). Extensive research has been done on the composition and size of groups to suit different tasks. In this case Joseph nominated the membership to form small groups of mixed ability and known group dynamics.

Joseph assigns roles to group members, and sets up monitoring systems. Research suggests a number of different roles that can be defined (Johnson and Johnson, 1990).

An STSP approach pushes into other areas of the curriculum. We need to be aware of the complexity of language tasks when we ask students to 'conduct an interview' or 'role play a debate'.

pin role descriptions on the notice board.

For this sequence, Joseph slows the pace to allow thinking and talking, and asks probing questions himself (Tobin and Fraser, 1988).

The first task is to develop questions (for the interviews) and checklists (for the observations). The data collection is a sophisticated language task: the children have to clarify their questions and write them, plan ways to record the data at the site, take notes and summarize main ideas. The children do the work in the language session, rather than science — it strengthens the links in the curriculum, but emphasizes the language aspects of the task.

Joseph asks each of the small expert groups to develop five questions, after which parallel groups come together to produce a single set. As they work, Joseph moves around, talking to them about how to write good questions, and listening to their ideas about things to watch for that make equipment strong. Joseph is happy to let this part of the unit proceed slowly so that the children can think through their own ideas, and develop relationships in their groups.

Joseph anticipates that the children will have difficulties with decision-making when they come together in groups of nine. To avert this, he 'cross pollinates' the small groups, and at the same time addresses directly the skills of working together.

To help the children reflect on their skills in cooperative work, he gives each group a ball of yarn and instructs the children that the ball should be always with the person doing the talking. When there is a change, the first child wraps the yarn around his/her wrist then passes the ball. At the end of the game, the web of yarn shows who has talked most often, and the paths of interaction. Joseph and the class discuss the various webs and what they mean, and

Learning to work in a group is not simply left to experience, but addressed directly — the functioning of groups is discussed, activities are appropriately sequenced and planned, special activities such as the web game (Graves and Graves, 1990) are included, and student progress is assessed.

the responsibilities of involving and participating.

When the groups of nine are formed, children retain the manager roles they had in the small groups. Each big group appoints a chairperson from the task managers. Joseph talks with the children about the chairperson role.

In a couple of instances, the groups can't agree on whether a question should be included. Joseph suggests that they allow the contentious questions to be asked in addition to the common set.

The children go back into their small groups to prepare their observation or interview charts for recording the data. Joseph makes notes on a checklist of his own. In some cases he chooses to be on hand to help.

A checklist focuses observation, enables a systematic approach to observation, and provides a record for later use and for reporting. (Lovitt and Clarke, 1988).

In the playground the activities go well. The Year 6 children see it as a culmination of their planning; the Year 1 children and the parents are happy to be involved. It is a sunny day.

Deciding What Pieces to Build

Back in the classroom, the children in their groups of nine compile and interpret their data — counting, comparing, summarizing. Differences between the playgrounds mean that the data collected by parallel groups are not entirely comparable and can't simply be pooled. As well, some students have never been to one or another of the playgrounds. Joseph is reluctant to take the group out again, so the onus falls to the children to describe equipment to one another. Joseph encourages them to make drawings and be guided by the Structures groups.

Joseph's plan overlooked the variations between playgrounds; in wanting to be scientific, he had been too focussed on the idea that the same data should be taken at the various sites.

They summarize their data on posters, using drawings and tables as well as sentences. The three posters will be photoreduced for the school bulletin. From each group the recorders report to the whole class, working from their posters. After the presentations, Joseph leads a class discussion to draw out generalizations.

'Going public' with the project and its progress is not only a responsibility to parents, but a means of boosting the students' self esteem and building their commitment (Malcolm, 1992a).

Some children, like Karen, are concerned for a conflict they see between safety and fun — the idea that equipment has to be dangerous to be fun. They use the class's data to support their claim, though they admit they have no good index of danger. Joseph asks the children to think about other instances where danger and/or violence seem to be part of the 'fun', especially in sports and films. Joseph suggests that the 'right' mix of danger and safety is an individual preference. He invites them to think about the ways different members of their family respond to dangers like heights, crossing busy roads, swimming in rough seas, riding roller coasters. Some children argue that the roller coaster is not really dangerous, but only imagined to be. Joseph asks the children to write their own views of the best mix of safety and danger in playground equipment, using examples, and ways that 'real danger' can be reduced while 'imagined danger' is maintained.

Any technological development has costs and benefits — for various groups, for the environment, for the short term and long term. In this instance, Joseph is able to draw attention to perceived value as against 'actual' value (USE, 1993).

The teacher's capacity to recognize and capitalize on a learning opportunity depends on the teacher's knowledge of substantive content, and the things children say as well as teaching methods. (Ingvarson, 1992).

It is time to decide which pieces to build. Joseph follows his plan (figure 5.1), postponing decisions about the final pieces until the children see how particular pieces fit together in the space. The children develop a list of preferred pieces based on their data and personal opinions. Joseph outlines the way the development of the submission will be managed — children will work in pairs to design and build models of chosen pieces, and then the class will

Again the focus is on management procedures, encouraging the children to think them through, criticize them, revise them, own them.

explore combinations and develop perhaps three or four alternative submissions. One student is concerned that this strategy denies integrated designs. The children talk about the integrated designs that they have seen and which ones are effective. They add some of these to the list on the board, as alternatives for the choosing. The children select which pieces they want to make and partners to work with.

All the pieces have to be made to the same scale. The children have their measurements of the actual equipment. They measure the drinking straws, satay sticks, pipe cleaners that they will use in the models. A scale of 1/10 means a climbing frame is one straw high, but some of the small rocking equipment will be hard to make on that scale. They decide to use 1/5. Using calculators, they set up a five point ready reckoner in a corner of the blackboard.

This is essentially mathematics — measuring, scaling, using calculators, set into the context of the project and student involvement in planning and management (Davidson, 1990).

The sense of community in the classroom is high and children are confident enough of each other that planning in a whole group discussion is successful.

To make large equipment, they will have to join straws. The children think up ways to make strong joins, such as putting folded pipe cleaners or rolls of paper inside the straws and reinforcing outside with sticky tape.

Some students feel that drinking straws are inadequate. Their strength won't compare with poles, especially if the straws have to be joined. Straws are flimsy whereas poles are heavy (although some equipment is made from plastic pipes).

Preliminary Design

The children are impatient to start building, but Joseph holds back, insisting on preliminary design work. He wants them to plan first on paper. To help their frustration, he arranges for an engineer to come and talk to them.

With 'design-make-appraise' activities in technology education (AEC, 1992), it is often difficult to get children to plan on paper. Yet this step is essential. It is during the planning phase, if it is done carefully, that creativity emerges and

He is able to enlist one of parents who is an engineer to organize a colleague to drop in for thirty minutes and talk about how she works. She brings with her some preliminary sketches and technical drawings, as well as photos of work in progress.

ownership of the designs develops (Malcolm, 1992a).

Joseph presents a female engineer to the students, to challenge the stereotypes and provide a role model for the girls in the class (Harding, 1986). He takes care to brief the woman, so she knows clearly what is expected of her and what the children are doing.

Joseph asks the children to compare the steps the engineer talked about in her design work with the steps the class is taking. Working in pairs, the children draw up two columns, and try to write down and compare the steps in both cases. Then they draw concept maps showing 'Things you do when you are designing'. Joseph talks to some of the children about their maps, and discusses with the class ideas of 'being scientific'.

A major aim of the unit is to help students understand the principles of design. As well as working through the process themselves the children reflect on the process and try to articulate its characteristics. This is difficult for the children. The lists and the concept maps are a basis for discussion.

They turn to the design task. Joseph explains the importance of developing options rather than seizing on your first idea. He likens the process to brainstorming. Working in pairs, they create three options and make for each a list of good and bad points. As they work, Joseph moves among the groups, looking especially for clues about their understandings of stability and strength. Joseph asks them to identify ways their designs meet the needs for balance (won't tip), rigidity (won't wobble) and strength (won't buckle). He encourages them to talk in their pairs and write down reasons for their claims. Joseph keeps close to children whom he identified as having trouble with their first drawings.

Research in management and decision-making as well as design show the importance of generating options and evaluating them (de Bono, 1985). de Bono notes that highly intelligent children often avoid generating options, preferring to put their energy to rationalizing and defending their first idea.

The groups report to a class discussion. Joseph takes care to involve early the students who seemed confused. Many children have a good understanding of the rules for balance and beam shapes. He allows their ideas to come up

The session focuses the class on explanations and theories, and makes public the range of ideas children have. It raises options for all children and gives them confidence for the next step, their investigations.

without judgment, much as in a brainstorming session, because he is preparing the children to design investigations of strength and balance.

He suggests investigations in three areas:

- Balance (including cantilevers, foundations and situations when children hang out of the equipment)
- Shape and rigidity (triangles, struts and strainers)
- Beams (flat beams, tubes, cylinders, H beams . . .)

He uses the same jigsaw groups and process as in the data collection exercise, except that he rotates the roles of Records Manager, Team Manager, and Task Manager.

The Jigsaw strategy is widely known (Dalton, 1985). It is an effective way of handling 'options' and parallel developments, but making tall ideas accessible to all students. As well, it requires children to articulate their own understanding as they teach other children in their home group.

As is his custom, Joseph takes care that the children know where they are headed and what is expected of them (Gunstone, chapter 10 in this volume).

The investigation is to test ideas and explanations listed on the board. He makes it clear that the tests should be specially designed and need not refer directly to the play equipment. He forewarns the children that after their investigations each group will demonstrate to the rest of the class what they have found out.

Joseph gives no particular directions to the Balance and Shapes groups, but prepares a worksheet for the Beams groups, to help them keep pace with the others. The sheet describes how various shaped beams can be made from newspaper, and alerts the children that the requirements of strength and weight might compete. Mid-way through the investigation, Joseph interrupts the groups so they can report to their home groups. Children explain the idea they are testing and how they are testing it, and invite comment. The discussion focuses on planning, not results, in an attempt to build students' skills in inventing strategies and thinking through 'what will happen if . . .'

A major objective of the unit is to help students understand design and develop their skills in design. They planned the project and along the way they designed interview schedules, playground equipment, and experiments to test structures. They also designed posters, demonstrations and reports. The extent to which skills in design and problem solving are transferable across tasks is not clear (Erickson, chapter 6 in this volume).

When the investigations are finished, Joseph helps the groups prepare brief demonstrations for the rest of the class. By working with the groups he minimises duplication as well as helping with preparation.

'Predict, explore, explain' has been used increasingly widely over the last decade (Baird and Mitchell, 1986; Gunstone, chapter 10 in this volume). It is a powerful way of eliciting and challenging children's conceptions. It can be led by students rather than teachers.

Some of the demonstrations take the form of 'Predict, observe, explain', where the child first asks the class to write down their predictions of 'what will happen when . . .' then observe the demonstration, and finally explain what happened.

Joseph gives out a work sheet on which all children write down their understandings of the rules of balance, rigidity and beam shape, and try to explain the rules. He collects the worksheets, and goes through them.

The unit contains a rich diversity of ideas and learning opportunities. Joseph wants to focus on understanding stability and strength. The worksheets serve to do this as well as provide Joseph with details on children's achievements on the objective.

Most students talk in terms of forces, squashing, weak points, and ideas of tipping over. Almost everyone seems to understand balance, and, implicitly, ideas of centre of mass and rotation (though they don't use those words). Some students can explain why bolting down a cantilever, setting the legs in concrete or making a heavy base prevents toppling. Most are confused however about struts and strainers and why some beams are stronger than others.

Joseph is unsure of how far to go. Few students can explain why the rules work. Many don't see any need to — their needs for explanation are satisfied by the rules. Joseph decides to do a little more work on struts, but leave beam shapes and girders for groups who finish their model building early.

The idea of 'levels of explanation' is important — what constitutes 'explanation' depends on the child's needs and knowledge as well as the teacher's and/or the subject (Havis, Malcolm and Seddon, 1992; Viennot, chapter 17 in this volume).

He chooses to give a demonstration, as the children did, building on their understanding of balance and cantilevers. He makes a cantilever and base from a

When children spend much of their time working in cooperative groups and projects, a break into traditional teaching is welcomed. Joseph's demonstration

piece of wood screwed to two supports. The children see that, as 'a person' walks out along the beam, the centre of gravity moves out and the beam rotates. They see also that the person can be counterbalanced by a weight at the back and that a heavier base or deep foundations makes the equipment more stable. Joseph draws attention to the idea that the supporting post at the front is squashed, and the one at the back is stretched to resist the rotation — he wants to orient the children to an explanation in terms of deformation, not forces.

is well based in ideas the children have already worked with.

From there Joseph turns to frame shapes. He compares a simple square (made from straws) to one with a strut across the corner. He pushes along the top of the square frame. It goes out of shape, but with little squashing or stretching of any of the sides. With the trussed frame, deforming it squashes the strut (the children can see it buckle) or stretches it. He does the demonstration without comment and the children write down their explanations of the differences (figure 5.2).

Joseph's explanation depends on deformation rather than force. Children have difficulty with Newton's forces and third law (Viennot, chapter 17 in this volume; Gunstone, chapter 10 in this volume; Minstrell, 1982). And in some ways the deformation argument provides a deeper explanation. (It explains why a structure or support is able to 'push back' against the deforming force and so explains Newton's third law in this instance).

Joseph shows the children pictures of some of the world's biggest bridges and tallest towers, and they note the use of struts and triangles. He asks them to think of examples around their area. The children's ideas range from the expected (towers and bridges), to novel (supports for Real Estate Agents' signs in front gardens) to the amusing (Nick described the way his uncle sat on a fold-up golfing seat-spike). On one of their drawings, Joseph asks the children to show how the strut is able to resist movement of the top of the frame.

Having explored concepts of stability and strength in the context of the project, the class applies them to other contexts. Joseph uses 'biggest bridges' and 'tallest towers' because such feats of human achievement are especially interesting to children at this age, as they come to grips with the limits of human possibility (Egan, 1976).

Constructing the Models

The children, working in their pairs, revise their favoured plan in the light

Because the preliminary work has been so detailed, this phase is 'downhill'.

Figure 5.2: Children's explanations of why triangles are more rigid than squares (year 6, Ashburton Primary School)

'The square has four joins and they all bend the same way, but the triangle pulls on itself and will not go.'

'The square can move easily. The corners are weaker . . . But the triangle (long bit) bends and the short bit doesn't want to bend.'

'The triangle stays still because to push it over, the length of the sides has to change and they can't.'

'With the triangle you have to push/pull against one pole.'

'The pole down the middle stops the triangle from moving sideways.'

'The triangle bends when you push and pull, and the square wobbles . . .'

'The square has more sides to make more movement.'

'The joints are weaker, [and] the rectangle has more joints than the triangle.'

'The triangle is the stronger because it is more compact . . .'

of their investigations. They draw it to scale on a poster (including some of the dimensions). On their poster they also list selling points of the design, and give advice about the ways the 'real' piece should be made. Joseph guides the children on poster production and talks about their designs. He makes notes on a student comment sheet he has prepared for the purpose.

Joseph's major role is to make time, and facilitate and support the children in their work (Johnson and Johnson, 1990).

The children set to work with straws, pipe cleaners, paper and card, sticky tape, satay sticks, pins and glue to make their models. They talk with each other across groups about ways to make joins. They show off their structures to each other and to Joseph. One boy puts a plasticine child on his equipment, and others follow suit.

Putting the Pieces Together

Standing in the open in the playground area, distances are deceiving. Joseph

This activity is a variation on 'predict explore explain' — the children confront

challenges the children to mark out how much space their models will take up. To guide them, he asks them first to estimate the width of the garden path. They write down their estimates and discuss them, and find that their values vary by a factor of five! They make measurements and check. Then they measure the playground area and draw a map.

their own perceptions of distance and scale before they begin measuring (Lovitt and Clarke, 1988).

Back in the classroom, they begin to mark a scaled down map on the floor. But it is four metres across! They move to the school hall.

Once again, managing the task and the children are critical roles for the teacher. Joseph, as before, involves the children in the process.

At one end of the hall, they mark out four copies of the playground area, and try placing their models. The plasticine people prove useful in helping understand the space required. Joseph leads the children in thinking through a process to decide the layout. They start with pieces which, from their studies, have high popularity. They have three or four of each of these pieces (including the integrated designs). They set them out, first one per area, then adding another, looking at combinations, and adding again until all pieces are out.

Some students are disappointed. The models are not as attractive as they expected. They feel that the integrated designs make better use of the space, that the class made the wrong choice in deciding to first make pieces and then look at combinations. Joseph accepts responsibility for the decision, and suggests that they write that idea in to their submission to council. But other students involved in the conversation aren't convinced that the integrated designs are better — they might look better, but they don't work as well, because children get in each other's way.

The possibility of children (and the teacher!) being disappointed with the outcome of their work is always real, especially when they invest so much of themselves in the work. (We see this similarly if a student has a high sense of commitment to and quality in work he/she submits but we grade it 'D'. In our story, Joseph saw no point in grading the children's models).

Joseph and the group call it quits for the day.

Joseph had in mind that the class would evaluate traffic flow and safety with the various layouts. Now that seems unwise. The children see the layouts as essentially the end of the story. He decides instead on a simpler evaluation, in which the year 1 respond to the designs. That will complete the story for the year 1 children too.

Finally the class has to develop the submissions. That will be straightforward because of the documentation that has been kept. He will arrange a notice board for each submission. The children can display their posters and reports, from the initial 'Things to consider lists', through the data summaries, experimental investigations, sketches and designs to the responses of some of the year 1 children to the actual models.

He arrives ready to suggest his plans, even though science is not scheduled today. A group of the children who were disappointed yesterday greet him with an idea: they want to lay some wood shavings among the models, and mound them up in some places, and put a border around them so the models look more realistic. They are excited, and he starts to tell them about his plans for presenting the submissions. Parents will like to see the displays too . . .

For reporting the students' progress and achievements, Joseph has available

- *lists, posters, interview and observation plans, data summaries, drawings, reports and models*
- *notes he made on presentations and group work, sometimes in 'free comment' format, sometimes using checklists*
- *worksheets/tests that the children completed on their explanations of strength and stability.*

Discussion

I have chosen to leave the story unfinished. This is partly because it is neither a tale, nor a unit to be replicated in other classrooms. It is intended to express curriculum principles and explore complexities of curriculum design, teaching, and learning outcomes that seem to be corollaries of the principles. The story is unfinished also because in fact there are many stories — for each of the children, the teacher, and the class as a group; for readers, according to the various lenses, purposes and biases they apply as they read.

It is interesting to speculate, especially with the help of a group, on what might come next and the effects it would have. How important is it to the story that the school council respond positively to the children's efforts? What difference would it make to the curriculum principles, the learning outcomes for the children and for Joseph if the council dismissed the children's work, or took ten months to make a decision, or employed immediately an equipment firm whose first step was to meet the children?

What if the story took a new turn? What if Joseph gained promotion to another school only to find that curriculum there was based on tight schedules, rigid structures and tidy desks? What might be the effect on Joseph and his career? How should he respond — knuckle under, start a revolution? Alternatively, what if the children go to a high school next year where science is teacher and text centred with tests every week? Can the children start a revolution?

Suppose instead Joseph stays at his current school and teaches grade 6 again next year. How will he teach Structures and Stability this time? What can he borrow from last year's effort? What is a suitable alternative context? Indeed what about right now, this next week, these next few days. Should Joseph and the children embark immediately on another unit of the style of *Design a Playground*? Should they take a rest for a while? But how? By reverting to 'routine teaching'. Does that mean traditional teacher and text centred show and tell approaches?

The recent research into constructivist learning, STS-P, inclusivity, and the development of broad skills has jolted teacher-student interactions in the classroom, dominated science education conferences, and changed the complexion of state and national policies on teaching and learning in Australia as in other places. Of course there is still more to be done on children's conceptions and ways to organize teaching sequences. But we need to attend also to the contextual issues of school and educational management. For example, how do we link the experience that Joseph's children have had in year 6 with the more formal work on forces that they will do later in years 9, 11 and 12? Whose is the responsibility? To what extent should the teaching and learning style of Joseph's class in year 6 be continuous with and consistent with the style of the high school they go to next year? How do we achieve that? We don't know how to achieve curriculum continuity and progression. We are not clear what progression means when we think over years rather than days or weeks, and we don't know how to write it at the school or system level. (Current attempts at National Statements and Assessment by Education Authorities in UK, various states of USA, and Australia, struggle to avoid being bureaucratic, overly simplistic, and possibly retrograde in their influence. As well they are unlikely to last for the twelve or thirteen years that children are at school).

The excitement of our time is that research in constructivist learning and related areas has created so many opportunities for teachers, children, researchers and educational administrators. The energy committed to teaching and learning interactions needs to be paralleled by innovations in assessment, school organization, teacher development and system level curriculum management, if much of our achievement is not to be diluted or dissipated.

References

AEC. (1992). *The National Technology Curriculum Statement*. Melbourne, Australia: Australian Education Council.

Baird, J.R. and Mitchell, I.J. (Eds.). (1986). *Improving the Quality of Teaching and*

Learning: The PEEL Project. Melbourne, Australia: Faculty of Education, Monash University.

BAIRD, J.R. and NORTHFIELD, J.R. (1992). *Learning from the PEEL Experience.* Melbourne, Australia: Faculty of Education, Monash University.

DALTON, J. (1985). *Adventures in thinking*. Melbourne, Australia: Nelson.

DAVIDSON, N. (Ed.). (1990) *Cooperative Learning in Mathematics*. Menlo Park, CA: Addison, Wesley.

DE BONO, E. (1985). *Tactics*. London: Fontana Collins.

EGAN, K. (1976). *Educational Development.* New York: Oxford University Press.

FENSHAM, P. (1988). Approaches to the teaching of STS in science education. *International Journal of Science Education, 10*, pp. 346–56.

GRAVES, N. and GRAVES, T. (1990). *A Part to Play*. Glen Waverley, Victoria, Australia: Latitude Publications.

HARDING, J. (Ed.). (1986). *Perspectives on Gender and Science*. London: Falmer Press.

HAVIS, S., MALCOLM, C. and SEDDON, J. (1992). *Electricity, STAV Physics Series*. Melbourne, Australia: STAV Publishing.

INGVARSON, L. (1992). *Professional standards for the teaching of science*. Unpublished report of the Science Education Professional Development Project, Canberra, Australia: Commonwealth Department of Employment Education and Training.

JOHNSON, D.W. and JOHNSON, R.T. (1990). Using cooperative learning in math. In DAVIDSON, N. (Ed.). (1990). *Cooperative Learning in Mathematics*. Menlo Park, CA: Addison, Wesley, pp. 103–125.

LOTAN R. and BENTON, J. (1990). Finding out about complex instruction: teaching math and science in heterogeneous classrooms. In DAVIDSON, N. (Ed.). (1990). *Cooperative Learning in Mathematics*. Menlo Park, CA: Addison, Wesley, pp. 47–69.

LOVITT, C. and CLARKE, D. (1988). *Assessment Alternatives in Mathematics*. Part of the Mathematics Curriculum and Teaching Program. Canberra, Australia: Curriculum Development Centre.

MALCOLM, C. (1987). *The Science Framework P-10: Science for Every Student*. Melbourne, Australia: Ministry of Education, Victorian Government,

MALCOLM, C. (1992a). *Science Teaching and Technology*. Carlton, Australia: Curriculum Corporation.

MALCOLM, C. (1992b). Science and personal development. *Australian Science Teachers Journal, 38*, pp. 8–14.

MINISTRY OF EDUCATION (1988). *The School Curriculum and Organisation Framework.* Melbourne: Ministry of Education, Victorian Government.

MINSTRELL, J. (1982). Explaining the 'at rest' condition of an object. *The Physics Teacher, 20*, pp. 10–14.

NCTM. (1991). *Professional Standards for Teaching Mathematics*. Virginia, USA: National Council of Teachers of Mathematics.

OSBORNE, R. and FREYBERG, P. (1985). *Learning in Science*. Auckland, New Zealand: Heinemann.

TOBIN, K.G. and FRASER, B.J. (1988). Investigations of exemplary practice in high school science and mathematics. *Australian Journal of Education, 32*, pp. 75–94.

UNESCO (1991). *Science for All and the Quality of Life*. Bangkok, Thailand: Regional Office for Asia and the Pacific.

USE (1993). *RISKO: The Game of Life*. Melbourne: Understanding Science and the Environment.

VICTORIAN CURRICULUM AND ASSESSMENT BOARD (1992). *VCE Implementation Handbook.* Melbourne, Australia: VCAB.

Chapter 6

Pupils' Understanding of Magnetism in a Practical Assessment Context: The Relationship Between Content, Process and Progression

Gaalen Erickson

There has been a continual debate in the science education literature for at least the last 100 years over the value of engaging pupils in practical, laboratory-type activities (Jenkins, 1989; Layton, 1973 and 1991). An integral part of this debate is the ongoing controversy over first, the extent to which pupils' explorations with materials leads to an increase in their understanding of scientific knowledge, and second, how this learning might occur. In this chapter I shall use the term 'practical work' to refer to a broad range of educational practices involving these explorations with materials. This range includes both general, instructional approaches and more specific teaching practices. Examples of the former would be the materials-based, instructional approaches in the primary grades (for example, the Elementary Science Study in North America and Nuffield Junior Science in England) or the laboratory-based activities in the secondary grades (Fensham, 1990). An example of the latter is the development of specific instruments to assess pupils' 'practical' knowledge and skills (Champagne and Newell, 1992; Giddings, Hofstein and Lunetta, 1991; and Kulm and Malcolm, 1991). While some have argued that engaging pupils in activities involving materials provides a basis for developing pupils' understanding of the so-called science 'process skills' (Wray, 1987), others have claimed that it provides the context in which pupils are able to construct a better understanding of scientific content (Brook, Driver and Johnston, 1989; Gunstone, 1991; White, 1988). I would submit that much of the substance of the contemporary debate regarding practical work and pupil outcomes can be summarized in terms of two general issues, expressed as questions. These are: (i) What is the nature of the relationship between pupils' understanding of science and their engagement in practical work? and (ii) Can we develop better or more 'authentic' (Wiggins, 1989) methods of assessing pupils' understanding of science? While I will primarily be addressing the first of these questions in this chapter, the data that I will be using to illuminate some of the issues associated with this question comes from a project designed to address the second question.

Some of the difficulty in addressing the role of practical work in science instruction lies in the inherent complexity and ambiguity of what it means to understand science. Much of this ambiguity comes from alternative ways to think about and represent scientific knowledge. Most of the participants in the past and present debates recognize that one can talk sensibly about both the methods and the content of science (see Easley, 1958, and Schwab, 1962, for earlier insightful analyses of this issue). The current controversy focuses around whether it is desirable or even possible to develop instructional programs (and assessment procedures) to teach (and measure) these two aspects of scientific knowledge as though they were *separate* cognitive outcomes (Hodson, 1988; Millar, 1991; Millar and Driver, 1987; Solomon, 1988; Woolnough, 1991). The dilemmas created by this dichotomy between method and content in the representation of scientific knowledge and the role played by practical work in the development of pupils' scientific understanding provide the two overarching themes for this chapter.

Purpose

The specific purpose of this chapter is to explore in greater detail the nature of this relationship between practical work and learning science by analyzing some recent data obtained from pupils in three different age groups when they participated in an open-ended investigation as part of a large-scale science assessment programme. These data do not permit an examination of any direct relationship between participating in practical activities and increased understanding; rather, they allow us to look at the relationship from another direction. The analysis is limited to two questions that examine the general issue of the role of practical work in learning science. These questions are:

1. What is the nature of the pupils' understanding of scientific knowledge that is being assessed by an open-ended, assessment task?
2. Is there any evidence of developmental differences or progression in pupils' understanding of science as indicated by the ways in which they approach and solve open-ended investigations in an assessment setting?

Context of the Assessment Project

The context, which generated the data presented in this chapter, was a large scale assessment project where we were developing performance tasks for the British Columbia Science Assessment Project (Erickson, Bartley, Carlisle, Meyer and Stavy, 1992).

While two types of performance assessment tasks were developed, only the 'investigation' task will be discussed in this chapter. This task involved two pupils from the same grade being presented with a set of materials and asked to answer an open-ended question given to them by the task administrator. They had up to sixty minutes in which to plan and perform one or more experiments to answer this question. Two of these tasks were designed to be used in an identical manner at three age levels — grades 4, 7, and 10 — but only the 'Magnets' task will be described here.

The Magnets Task

This task presented the pupils with three magnets which were similar in size, mass and appearance (except that they were painted different colours) and a set of materials. The question that the pupils were encouraged to answer was: 'Which magnet is the strongest?' The materials used in this task were identical for each of the three grade levels. For a more detailed description of this task and the materials used see Erickson *et al.* (1992).

The Pupils

There were approximately 120 pupils completing the magnets investigation task at each of the three grade levels. These pupils were selected from classrooms in randomly selected schools throughout the Province of British Columbia. In total, over all three grade levels, over eighty schools were involved from eighteen different school districts. The pupils worked in pairs and they were encouraged to work together and to discuss their ideas with each other as they worked with the materials.

The Teacher-administrators

As part of the overall performance assessment project a decision was made to provide classroom teachers with the necessary experience and knowledge to administer and analyze all of the assessment tasks. Thirty-six teachers from the eighteen school districts were engaged in a three-day, orientation workshop where they were introduced to the various tasks, worked with groups of pupils, practiced the set-up procedures, and discussed the purpose and rationale of these assessment tasks. After these teachers administered the tasks in their own school districts they participated in a further three day coding and analysis workshop where all of the pupil responses were coded.

Administering and Coding the Tasks

Once the teacher-administrator (hereafter referred to simply as the teacher) was introduced to the two pupils selected to do the investigation, they all sat down at a comfortable working space with the teacher sitting across from the pupils. The purpose of the activity was explained to the pupils along with a quick introduction to the materials available for the investigation — identifying the names of any materials that the pupils did not recognize. The teacher completed an observation schedule and made notes while observing the pupils work.

These observation schedules, along with any notes that the teachers used to supplement their descriptions of the pupils' performance on this task, were used to complete a coding sheet during the three-day coding and analysis workshop. The coding sheet follows the structural characteristics of the observation schedule fairly closely. An example of the Magnets coding sheet is given in Figure 6.1. The teachers were asked to select what they considered to be the 'best experiment' of all those attempted by each pair and to complete a set of four judgments about this one experiment. Finally, they were then asked to give a global rating, on the same five point scale used for the earlier judgments, of the pupils' performance considering all of the experiments that they had attempted. This Judgment Sheet is shown in figure 6.2. An important caveat must be made on the issue of the comparability of judgments across the grades. Different teams of teachers were used at the three grade levels and there was not sufficient time during the coding sessions for the three teams of teachers to meet together to determine whether it was either possible or desirable to try and produce a common set of criteria across all three grade levels for these investigations tasks.

Results of the Magnets Investigation

Once all of the coding sheets were completed, the data were then tabulated and a variety of frequency displays were generated. While we have data on all of the experiments performed by the pupils, only the data for the experiments judged by the teacher to be the pair's 'best experiment' are presented below.

The teachers were asked to code both the general approach that the pupils seemed to be using and the 'implicit questions' that appeared to be guiding their actions. We refer to these as 'implicit' because in most instances the pupils did not identify (either verbally or in writing) this question. Since the judgments about the 'approaches' and the 'implicit questions' are connected, as shown in figure 6.1, only the data on the pupils' implicit questions are given in table 6.1. These data provide some interesting insights into pupils explorations with magnets. Virtually all of the pupils at all three grade levels, who used a 'distance approach', were concerned with determining the distance between an object (for example, a steel ball) and a magnet just before it was attracted or pulled to the magnet. Two pairs of grade 7 pupils were the only exception to this strategy as they used iron filings to try to determine the range of the influence of the magnet of the filings.

When we examine the breakdown of those implicit questions that were coded under the 'number of objects' approach, we see that almost half (43 per cent) of the grade 4 pupils seemed to be asking the question of 'How many objects can each magnet hold?' Whereas the older pupils were more inclined to be asking questions about the number of objects that can be 'chained', or 'held through a barrier'. There does appear to be a relationship between pupil familiarity with materials and their approach to the problem. An illustration of this point is the fact that no grade 4 pupils opted to use a spring balance (not something that most primary classrooms would have available) while 14 per cent of the grade 7 pupils used this more precise strategy for measuring the force that a magnet exerts on an object.

To obtain an even more detailed examination of how the pupils engaged with this problem setting, data were coded on: whether the pupils appeared to be systematically controlling the variables that they were manipulating and measuring, the nature of the measurement strategy they were using, and the final outcome of their experiment. These data are given in table 6.2. One must be cautious in interpreting aggregated data of this nature since it is not possible to match

Figure 6.1: Coding sheet for magnets investigation

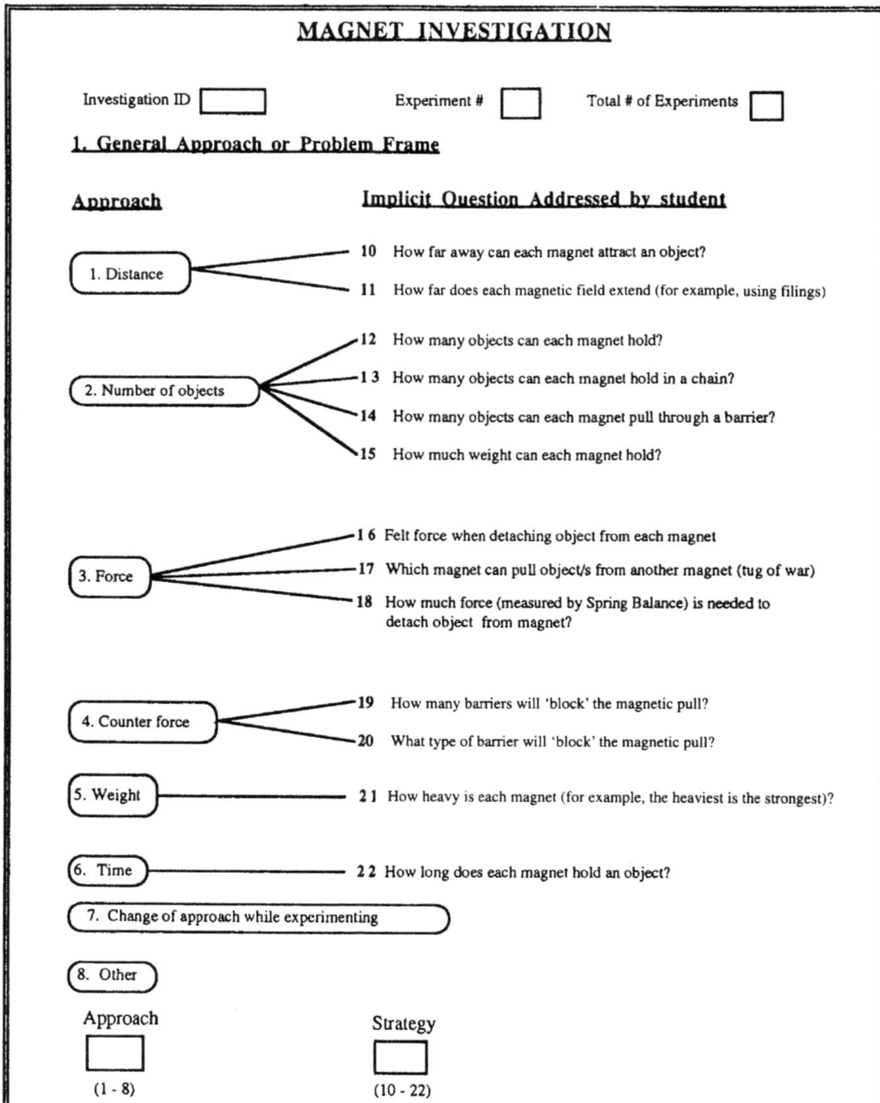

MAGNET INVESTIGATION

Investigation ID ☐ Experiment # ☐ Total # of Experiments ☐

1. General Approach or Problem Frame

Approach — **Implicit Question Addressed by student**

1. Distance
- 10 How far away can each magnet attract an object?
- 11 How far does each magnetic field extend (for example, using filings)

2. Number of objects
- 12 How many objects can each magnet hold?
- 13 How many objects can each magnet hold in a chain?
- 14 How many objects can each magnet pull through a barrier?
- 15 How much weight can each magnet hold?

3. Force
- 16 Felt force when detaching object from each magnet
- 17 Which magnet can pull object/s from another magnet (tug of war)
- 18 How much force (measured by Spring Balance) is needed to detach object from magnet?

4. Counter force
- 19 How many barriers will 'block' the magnetic pull?
- 20 What type of barrier will 'block' the magnetic pull?

5. Weight
- 21 How heavy is each magnet (for example, the heaviest is the strongest)?

6. Time
- 22 How long does each magnet hold an object?

7. Change of approach while experimenting

8. Other

Approach ☐ (1 - 8) Strategy ☐ (10 - 22)

Figure 6.1: (cont.)

2. Performance

What is Controlled For?

3 0 same size and type of objects
3 1 same number of objects
3 2 same side of magnet
3 3 same number of barriers
3 4 same type of barriers
3 5 other

Procedures

4 0 one magnet at a time
4 1 two magnets at a time
42 two or more magnets in competition
4 3 parallel independent experiments by each individual
4 4 parallel cooperative experiment

Measurement Types

5 0 counting
5 1 linear measurement of distance
5 2 weighing
5 3 estimating
5 4 computing

Appropriateness of Control

6 0. Control is appropriate for the design selected
6 1. Control is not appropriate for the design selected.

3. Outcomes

70 Black
71 Blue
72 Red
73 Silver
74 Tie between two magnets
75 Abandoned experiment
76 Inconclusive results
77 Other

Figure 6.2: Judgment sheet for magnets

JUDGMENTS OF STUDENTS' BEST EXPERIMENT

Investigation ID ☐ Best Experiment # ☐ Total # of Experiments ☐

Use this scale to code your judgments about the students' best performance for each of the following questions:

poor		satisfactory		excellent
1	2	3	4	5

1. How well do the students plan an experiment to answer the question? ☐

2. How well do the students develop a suitable measuring strategy? ☐

3. How well do the students interpret the data collected to answer the question? ☐

4. How well do the students report the results of their experiment? ☐

JUDGMENT OF STUDENTS' OVERALL PERFORMANCE

Use the same scale to evaluate the students' performance over all their experiments.

5. In your judgment, considering **ALL of the students' experiments**, how would you rate their performance on this task? ☐

these data directly to the earlier table on implicit questions. However, this table does identify some interesting features of the pupils' investigative abilities. One significant feature of the data in section A of table 6.2 is that the grade 4 pairs appear to be very proficient at designing experiments where the variables of 'size and type of object' are controlled. These are the variables that are relevant to the questions of: 'How many objects can a magnet hold?' and 'How far away can a

Table 6.1: Implicit question addressed by student for the magnets task[1]

	Gr. 4 (%)	Gr. 7 (%)	Gr. 10 (%)
How far away can each magnet attract an object?	23	24	34
How far does each magnetic field extend (for example, using filings)?	0	4	0
How many objects can each magnet hold?	43	12	13
How many objects can each magnet hold in a chain?	13	8	20
How many objects can each magnet pull through a barrier?	7	16	7
How much weight can each magnet hold?	0	8.	2
Which magnet can pull object/s from another magnet (tug of war)?	9	6	5
How much 'felt force' is required to detach object from magnet?	0	4	2
How much force (measured by spring balance) is needed to detach object from magnet?	0	14	9
How many barriers will 'block' the magnetic pull?	0	2	4
What type of barrier will 'block' the magnetic pull?	4	2	2
How heavy is each magnet (for example, the heaviest is the strongest)?	0	0	2
How long does each magnet hold an object?	2	0	0

[1] These data are only for the pupils' best experiment.

Table 6.2: Some features of students' experiments with magnets

A Controls			
Controlled for	Gr. 4	Gr. 7	Gr. 10
1. same size and type of object	89[1]	92	87
2. same number of objects	70	80	54
3. same side of magnet	20	69	75
4. same number of barriers	4	16	11
5. same type of barrier	5	10	13
6. other	14	2	7

B Measurement	Gr. 4	Gr. 7	Gr. 10
Measured			
counting	55[1]	49	40
linear measure of distance	15	36	33
weighing	4	8	13
estimating	38	12	15
computing	0	2	0
not appropriate	5	8	2

C Outcomes	Gr. 4 (%)	Gr. 7 (%)	Gr. 10 (%)
black	65	93	96
blue	11	0	0
silver	19	2	0
tie	4	2	2
abandoned	2	4	0
inconclusive	0	0	2

[1] These figures represent the total number of entries on the coding sheets. Up to three entries for each 'best experiment' were permitted for the variables controlled and two entries for the type of measurement used.

magnet attract an object?' In other words, these are the most appropriate variables to control for the implicit questions the grade 4 pupils seemed to be addressing. This result certainly challenges some of the conventional wisdom in the field of science education which claims that pupils experience difficulty in controlling variables until they reach Piaget's stage of formal operational thought — around the age of 13 to 14.

Likewise for the pupils in grade 7 and 10 the data certainly seem to indicate that these pupils were very competent at designing experiments with appropriate controls and measurement strategies and subsequently arriving at a justifiable conclusion. Well over 90 per cent of these pairs correctly identified the black magnet as being the strongest. In examining the data in table 6.2 for differences in performance between the grade levels, several important issues emerge. First, it is noteworthy that only twenty instances of controlling the variable 'the same side of the magnet' occurred at the grade 4 level while sixty-nine and seventy-five occurred at the grade 7 and 10 levels respectively. This finding (along with accompanying anecdotal comments from the teachers) suggests that the grade 4 pupils are not aware of the polarity of the faces on this type of ceramic magnet and hence may have used different faces for each of the three magnets they used in their experiments. Failure to use the same face for each magnet could certainly result in ambiguous results and could account for the 19 per cent of the pupils who found the silver magnet to be the strongest. The critical point following from this discussion is the extent to which the pupils' background knowledge influences both the way in which an experiment is framed and performed. If pupils do not know that a magnet has poles, or do not recognize these poles for a given magnet, then clearly they will not worry about controlling for the face of the magnet they are using.

Another important difference between the pupils at the three grade levels is found in section B of table 6.2 on the measuring strategies. Many more of the grade 4 pupils used estimation as a means of measurement than did the grade 7 or 10 pupils. This could easily lead to difficulties in distinguishing between the strengths of the black and silver magnets, or in estimating the number of objects attached to a magnet (especially in the approach used by some of the pupils where they piled as many objects on the magnet as it would hold).

In summary, table 6.2 provides a very encouraging picture of pupils' ability to work with these materials in an intelligent and mature manner in order to arrive at a reasonable solution to the question they were given at the beginning of the investigation. A further source of data to verify this optimistic view of pupils' investigative abilities comes from the teachers' judgment sheets (see figure 6.2). Table 6.3 presents the aggregated data for all of these judgments. As indicated earlier, the judgments across the three grade levels cannot be compared directly since they were made by different groups of teachers. If we aggregate the teachers' judgments to include the satisfactory to excellent ratings (creating a degree of satisfaction measure), we obtain some rather outstanding results for the pupils' performance on the magnets task. At the grade 4 level most of the judgmental categories, with the exception of reporting results (72 per cent), have a satisfactory rating of around 80 per cent. Similarly for the grade 7 pupils these rating are well over 90 per cent on most of the categories; again the only exception being 'reporting results' which was 73 per cent. The grade 10 ratings are consistently higher than even the other two grades with the satisfaction measure

Table 6.3: Teachers' judgments for the magnets investigation

Score	Plan (%)			Develop (%)			Interpret (%)			Report (%)			Overall (%)		
	4	7	10	4	7	10	4	7	10	4	7	10	4	7	10
1	5	2	0	11	2	0	14	2	0	15	4	0	5	6	0
2	14	0	6	11	8	2	5	4	4	13	23	5	14	2	4
3	33	17	38	23	13	20	21	9	21	24	21	29	26	28	31
4	26	23	40	35	17	46	26	8	32	22	26	29	21	40	49
5	21	59	16	21	60	32	33	77	43	27	26	38	33	25	16

exceeding 95 per cent on four out of the five categories. These consistently high ratings over all three grade levels are compelling evidence that the pupils certainly possess the abilities to plan and conduct small-scale investigations of this nature.

Pupils' Knowledge and Task Performance

Let me return to the first question posed earlier in this chapter: What sort of knowledge were the pupils displaying in the completion of this assessment task? The above data provide us with descriptions of pupils' activities and some insights into the various approaches and strategies that the pupils adopted as they worked with the magnets. But can we call this knowledge, and if so, what kind of knowledge is it? Typically this type of knowledge of pupils' actions in an inquiry setting has been referred to as procedural (or process-oriented) as opposed to propositional (or content-oriented). But this distinction between method and content, as was discussed above, has become increasingly blurry. For example, in the magnets investigation what is it that the pupils' *know*, as they try to determine the relative strength of the three magnets using a 'distance' approach? Are the approaches, strategies, and features of the pupils' experiments, categorized above, indicators of their investigative abilities (skill in methods)? Or are they simply general, cognitive abilities being brought to bear on a problem situation which requires an understanding of the properties of magnets, of the friction of rolling balls, of the importance of consistency of results obtained by repeating the procedure several times, of using a ruler to measure the distance, and so on? From the latter perspective the pupils appear to be drawing upon a large repertoire of different types or 'pieces' (diSessa, 1988) of knowledge as they become engaged in this assessment task. Furthermore, all of these 'pieces' of knowledge appear to be relevant in the pupils' planning and performing of this experiment. But some of these pieces are certainly more content-dependent than others and would require previous experience with magnets and rolling balls. Other pieces, such as the importance of the reproducibility of results and the use of appropriate measurement techniques, would appear to be more generic in nature and look more like the kinds of generic methods or scientific process skills sought by some (Bryce and Robertson, 1985; Wray, 1987). Other researchers have openly acknowledged the critical role of content knowledge in addressing problems like the magnets task, but argue that there are also more generalizable 'cognitive skills' (Perkins and Salomon, 1989) or 'scientific thinking skills' (Carey *et al.*, 1989; Kuhn, Amstel and O'Loughlin, 1988) that need to be integrated with the pupils' contextually-bound, content understanding of the phenomena under scrutiny.

Two recent literature reviews have addressed a number of the issues raised above, in particular the attempts to develop a conceptualization of knowledge that might lead to a better understanding of the relationships between domain-specific, content knowledge and generalizable process or strategic knowledge (Furnham, 1992; Layton, 1991). They develop the intellectual and practical contexts in which these problems reside as well as articulating the complexity of the issues in much greater detail than is possible here.

Another recent conceptualization of the specific issues being addressed by this chapter has been advanced by Millar and colleagues in a series of papers (Millar, 1987, 1989 and 1991; Millar and Driver, 1987; Millar and Wynne, 1988). In these papers they argue that it is 'misleading to portray the methods of science in terms of discrete processes', that many of the so-called processes are 'general cognitive skills which we all routinely employ throughout our lives, without any need for formal instruction', and that 'there is no evidence that we improve on our performance of any of these (content-independent) processes' (Millar, 1989, p. 49). Millar suggests that some of the confusion over the nature of practical work (and the knowledge that pupils construct from these experiences) might be reduced by distinguishing between three different types of 'practical skills' (Millar, 1991). The first such skill he calls 'general cognitive processes' and it includes such processes as: observing, classifying, inferring and hypothesizing. These skills, he claims, are essential to thinking in general and they are in evidence in very young children — perhaps even at birth (see Feldman, 1980, for a similar kind of argument where he claims that many of the basic concepts and cognitive processes being studied by Piaget in children's thinking up to about the age of 12 are 'universals', that is they are constructed by all children without the aid of formal instruction). The second category is 'practical techniques' by which he means knowledge about the appropriateness and the ability to carry out specific procedures such as measuring the property of an object (for example, temperature or mass) or separating different substances (for example, by filtration or chromatography). These techniques can, and must, be taught if we expect pupils to be able to use them in classroom or assessment settings. The final category he describes as 'inquiry tactics' which refer to the types of strategies used in producing a valid and reliable design for the experimental inquiry. Included in this category are such procedures as the identification and controlling of relevant variables, representing and transforming data in an appropriate ways to determine patterns and trends, taking repeated measures, etc. It is in this last category that content knowledge has the greatest interactive effect since tactics such as the selection and control of variables depends largely on the nature of the pupils' understanding of the phenomena being examined. For example, consider the conjecture advanced in the results section above to account for why many of the grade 4 pupils did not control the variable of 'the same face' of the magnet during their investigations (the conjectured explanation was that these pupils likely did not understand the polarity of magnets). As with the practical techniques, these tactics must be taught and it is in these last two categories that one would expect and hope to see some differences, or progression, occurring between pupils at the different grade levels.

Progression and Pupils' Understanding of Science

This leads to the second question, posed above, regarding the issue of whether any developmental trends, or progression, can be detected in the pupils' performance across the three grade levels. Is there any evidence of progression in the pupils' use of 'practical techniques' and 'inquiry tactics', as suggested above? The types of practical techniques that could be displayed in the magnet investigation are those associated with measuring the effects of the interaction between the magnets and the selected objects that were made available to the pupils. In looking at the data in table 6.2, section B on measurement, we see some indication that the grade 4 pupils tended to use somewhat less sophisticated measurement strategies than the grade 7 or 10 pupils. For instance they tended to rely much more frequently on estimating, they simply counted objects more often, and used a linear measure of distance less often. While some of these measuring strategies adopted by the grade 4 pupils may have been valid for the implicit question that they used to frame their experiment, it is also clear that some were not (for example, for many of the 23 per cent of the pupils who were determining the distance that a magnet can attract an object). Thus a case could be made that some progress seems to have been made in the development of more sophisticated measuring strategies, particularly between the grade 4 and the grade 7 pupils. Interestingly, there does not appear to be much in the way of differences in this domain of measurement techniques between the grade 7 and 10 pupils.

Are there any differences in the 'inquiry tactics' employed by the pupils at different age levels? The only inquiry tactic that we examined in some detail was the pupils' identification and control of variables that were judged to be appropriate for the implicit question(s) they were attempting to answer (i.e. the design of their experiment). As discussed above, these data are not easy to interpret because of the strong interaction between the pupils' understanding of magnetism and the implications of this understanding for the way the pupils subsequently framed their experiments.

To obtain further insight into this issue of the interaction between content knowledge and inquiry tactics, it may be useful to digress momentarily and turn to some work completed by Karen Meyer (1991) in her doctoral dissertation. She documented some of the explanatory models of magnetism that pupils aged 9 to 14 appeared to be using when placed in a similar problem-solving setting. These models are displayed in figure 6.3. The 'Pulling Model' must be considered to be a very primitive model in that it only provides a description and a naming of the process accounting for the pupils' observations (i.e. the magnet 'sucks the objects in'). However, for the majority of the grade 4 pupils in our sample this seemed to be their preferred way of thinking about magnets since the majority of the implicit questions that they used (for example, 'How many objects can each magnet hold?') would appear to be based upon such a conception. This conjecture is also supported by Meyer's findings based on several in-depth interviews with four pairs of pupils at each of the grade 4 and 7 levels.

Can we see any evidence of the other two models in the implicit questions framed by the pupils in our sample? It would seem that the distance strategies (for example, 'How far away does the magnet attract an object?') can be generated from either the 'Emanating Model' or the 'Enclosing Model'. As the diagram in figure 6.3 clearly indicates, one of the features that Meyer found in pupils who

Figure 6.3: Student models of magnetism

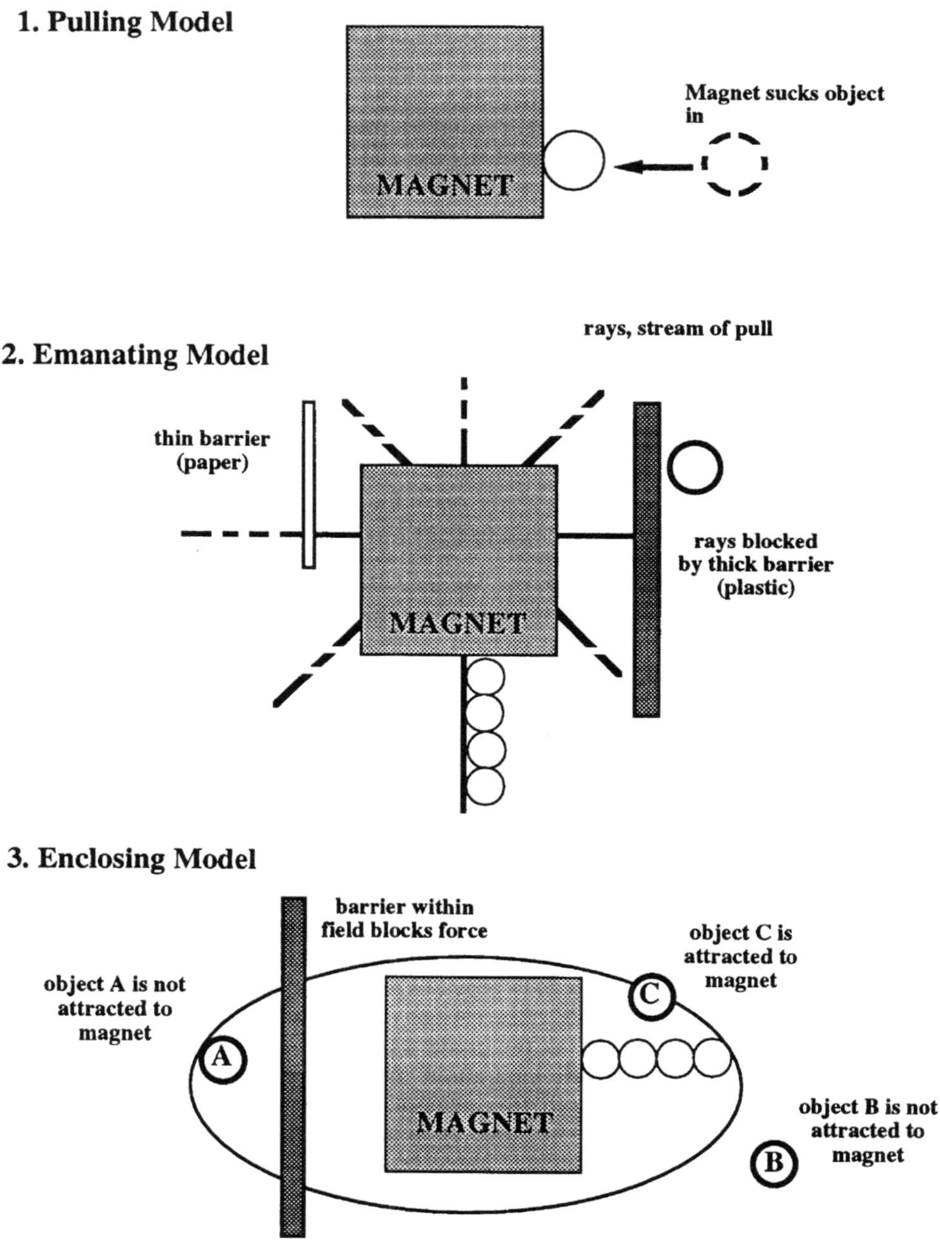

seemed to be appealing to an 'Enclosing Model' was the use of barriers to try and block the effects of the field generated by the magnet. Those questions concerning the effects of 'barriers blocking the magnetic pull' would seem to be more directly associated with this third model.

In returning now to the issue of progression, the question we must ask is: Do these three models represent a development trend or sequence? A second question follows: Is there any evidence that the older pupils approached the planning of their experiments using the kind of framing associated with one of the more 'progressive' models?

As suggested above, it seems clear that the 'Pulling Model' is fairly primitive in that it really does not posit any mechanism to account for the observed phenomena of magnetic attraction, other than attributing the somewhat animistic property to the magnet of having an inherent 'sucking' or 'pulling' capability. This is reminiscent of an early stage of causal reasoning described by Piaget (1969) as 'animistic causality' or 'dynamic causality'. The second and third models (the 'Emanating' and 'Enclosing' models respectively) have somewhat more explanatory power in that they postulate a simple mechanism for a *force acting at a distance* whereby 'rays' or some other subtle fluid emanate from the magnet (in the case of the 'Emanating Model') and interact with metal objects in such a way as to produce an attractive force. In the 'Enclosing Model' these rays seem to *dissipate* and so create a definite *area of influence*. Some of the pupils in Meyer's study referred to this area in terms of a magnetic field (borrowing the language of contemporary physics that they had no doubt heard or read about in school, in the media, or from family discussions). Both of these models take on most of the characteristics that Andersson (1986) describes as an 'experiential gestalt of causation'.

I would argue that these last two models represent an increase in explanatory power because they not only explain more of the phenomena that the pupils are encountering in this type of setting, but they also open up more possibilities for generating new experiments. Thus the pupils' general approaches which sought to determine the distance of influence that the magnet has, the use of different types of barriers to try and block the 'magnetic rays', and the use of the 'chaining' strategies with washers or steel balls all would appear to have been generated from the 'Emanating' and 'Enclosing' models.

Is there any evidence of progression in terms of the age of the pupils who seem to have subscribed to one of these more powerful models? The evidence on this issue must necessarily be highly inferential since we did not ask the teachers to collect any data on the pupils' ideas about magnets or magnetism; rather the conjectures are based upon a general coherence between the models proposed by Meyer and the 'implicit questions' that were coded and tabulated in table 6.1. The implicit question most closely associated with the 'Pulling Model' would seem to be the straightforward one of 'How many objects can each magnet hold?'. We see that 43 per cent of the grade 4 pupils were coded under this question, in comparison with only 12 per cent and 13 per cent of the grade 7 and 10 pupils respectively. It is somewhat more difficult to identify 'implicit questions' as belonging uniquely to either the 'Emanating Model' and the 'Enclosing Model' since the latter might well be, as suggested above, a slightly more developed version of the former. Thus it would appear that both of these models could be responsible for generating questions like 'How far away can each magnet attract an object?'; 'How many objects can each magnet hold in a chain?'; and 'How many and what type of barriers will *block* the magnet pull?'. Although there is considerable dispersion of these implicit questions among the pupils at all three grade levels, on balance it does seem that more grade 10 pupils opted for these latter types of questions than did the grade 4 or 7 pupils. In summary, these data would seem to support a claim that the older grade 7 and 10 pupils do indeed display a greater diversity and more sophisticated 'practical techniques' than the younger grade 4 pupils. Furthermore, the older pupils also appear to have constructed somewhat more powerful models of magnetism as inferred from the 'inquiry tactics' that they employed in this investigation.

Conclusions

The results obtained from this assessment project on first appearance left our project team somewhat puzzled. First, both the teachers administering and coding the tasks and we were surprised at how well the grade 4 pupils were able, not only to cope with the demands of the task, but to perform at a very acceptable level using many of the same strategies as those employed by the older pupils. This leads to a second point concerning our expectations; we thought that we would see much larger differences between the younger and the older pupils than we actually obtained. Can these results be explained in some manner? Some of these findings appear to 'fit' (von Glasersfeld, 1989) a perspective on learning that seems to be gaining increasing prominence. This general theory on cognitive development, or learning, is variously characterized as 'situated cognition', 'cognition in practice', or 'everyday cognition' (Brown, Collins and Duguid, 1989; Lave, 1988; Rogoff and Lave, 1984). Briefly put, this perspective claims that learning occurs in everyday situations where thinking is directed towards addressing a problem that is embedded in a practical context. Layton (1991) summarizes this view of cognitive growth as follows:

> In everyday situations where practical action is required, people devise satisfactory opportunistic solutions which, far from being illogical and sloppy, are sensible and effective in handling the practical problem. (p. 65)

Thus in the present chapter, the pupils have no doubt had a number of everyday experiences with magnets that permitted them to explore, manipulate, and so develop an understanding of the properties of magnets. Using their 'general cognitive skills' (Millar, 1991) pupils, by interacting with magnets and other materials, gradually construct intuitive 'explanatory models' like those outlined above. These models are of sufficient generality that they can be used in many situations that the pupils encounter in their 'everyday lives' and, in some instances, in their 'school lives' as well. But unless they are provided with the opportunities and the techniques to extend these models, their understanding of phenomena *and the role that their own construction of models to investigate those phenomena* will not progress beyond a level of common sense knowledge.

One of the responsibilities of science teaching, then, is to provide pupils with the types of practical activities that will enable the pupils to acquire a larger repertoire of 'practical techniques' and more powerful 'inquiry tactics'. But this repertoire of techniques and tactics by itself, is not sufficient if we wish pupils to construct more powerful and comprehensive models (of magnetism for instance) than those which seemed to shape the pupils' actions in our assessment project. In addition to acquiring these 'tools' of inquiry the pupils must be a part of a learning environment where they can participate in what has been described by some as a 'cognitive apprenticeship' (Brown *et al.*, 1989). This environment will entail creating the types of pedagogical structures where pupils and teacher work together to create a 'collaborative community' (Erickson, 1991; Granott, 1991) for the purpose of not only constructing more powerful models of the scientific content being considered (the underlying theme of many of the chapters in this volume), but also an increased understanding of their own learning processes (for example, as discussed in chapter 10 by Gunstone in this volume). The challenging task facing all educators, then, is to create the kinds of pedagogical contexts to support these collaborative communities using socially-relevant, content domains (such as those featured in the chapters in this volume: our bodies, our physical environments, our playgrounds, energy and chemical change in our everyday lives, etc.) that are stimulating, engaging, and enriching for *all of the participants* in this community.

References

ANDERSSON, B. (1986). The experiential gestalt of causation: A common core to pupils' preconceptions in science. *European Journal of Science Education*, *8*, pp. 155–71.

BROOK, A., DRIVER, R. and JOHNSTON, K. (1989). Learning processes in science: A classroom perspective. In WELLINGTON, J. (Ed.). *Skills and Processes in Science Education*. London: Routledge, pp. 62–82.

BROWN, J., COLLINS, A. and DUGUID, P. (1989). Situated cognition and the culture of learning. *Educational Researcher*, *18*(1) pp. 32–42.

BRYCE, T. and ROBERTSON, I. (1985). What can they do? A review of practical assessment in science. *Studies in Science Education*, *12*, pp. 1–24.

CAREY, S., EVANS, R., HONDA, M., JAY, E. and UNGER, C. (1989). 'An experiment is when you try it and see if it works': A study of grade 7 students' understanding of the construction of scientific knowledge. *International Journal of Science Education*, *11*, pp. 514–29.

CHAMPAGNE, A. and NEWELL, S. (1992). Directions for research and development: Alternative methods of assessing scientific literacy. *Journal of Research in Science Teaching*, *29*(8), pp. 841–60.

DISESSA, A. (1988). Knowledge in pieces. In FORMAN, G. and PUFALL, P. (Eds.). *Constructivism in the Computer Age* (pp. 49–70). Hillside, NJ: Lawrence Erlbaum Associates.

EASLEY, J. (1958). Is the teaching of scientific method a significant education objective? In SCHEFFLER, I. (Ed.). *Philosophy and Education* (pp. 154–177). Boston, MA: Allyn and Bacon.

ERICKSON, G. (1991). Collaborative inquiry and the development of science teachers. *The Journal of Educational Thought*, *25*(3), pp. 228–45.

ERICKSON, G., BARTLEY, T., CARLISLE, R., MEYER, K. and STAVY, R. (1992). *The 1991 British Columbia Assessment of Science, Part II: Student Performance Component.* Victoria, BC: Queen's Printer.

FELDMAN, D. (1980). *Beyond Universals in Cognitive Development.* Norwood, NJ: Ablex Publishing Corporation.
FENSHAM, P. (1990). Practical work and the laboratory in science for all. In HEGARTY-HAZEL, E. (Ed.). *The Student Laboratory and the Science Curriculum.* London: Routledge, pp. 291–311.
FURHAM, A. (1992). Lay understanding of science: Young people and adults' ideas of scientific concepts. *Studies in Science Education, 20*, pp. 29–64.
GIDDINGS, G., HOFSTEIN, A. and LUNETTA, V. (1991). Assessment and evaluation in the science laboratory. In WOOLNOUGH, B. (Ed.). *Practical Science.* Milton Keynes: Open University Press, pp. 167–78.
GRANOTT, N. (1991). Separate minds, joint effort and weird creatures: Patterns of interaction in the co-construction of knowledge. In WOZNIAK, R. and FISCHER, K. (Eds.). *Specific Environments: Thinking in Context.* Hillside, NJ: Lawrence Erlbaum Associates.
VON GLASERSFELD, E. (1989). Cognition, construction of knowledge, and teaching. *Synthese, 80*(1), pp. 121–40.
GOTT, R. and MASHITER, J. (1991). Practical work in science — a task-based approach? In WOOLNOUGH, B. (Ed.). *Practical Science.* Milton Keynes: Open University Press, pp. 53–66.
GUNSTONE, R. (1991). Reconstructing theory from practical experience. In WOOLNOUGH, B. (Ed.). *Practical Science.* Milton Keynes: Open University Press, pp. 67–77.
HARLEN, W. (1991). Performance testing and science education in England and Wales. In KULM, G. and MALCOLM, S. (Eds.). *Science Assessment in the Service of Reform.* Washington, DC: American Association for the Advancement of Science, pp. 163–185.
HODSON, D. (1988). Experiments in science and science teaching. *Educational Philosophy and Theory, 20*(2), pp. 53–66.
JENKINS, E. (1989). Processes in science education: An historical perspective. In WELLINGTON, J. (Ed.). *Skills and Processes in Science Education.* London: Routledge, pp. 21–46.
KUHN, D., AMSTEL, E. and O'LOUGHLIN, M. (1988). *The Development of Scientific Thinking skills.* New York: Academic Press.
KULM, G. and MALCOLM, S. (Eds.). (1991). *Science Assessment in the Service of Reform.* Washington, DC: American Association for the Advancement of Science.
LAVE, J. (1988). *Cognition in practice.* Boston, MA: Cambridge University Press.
LAYTON, D. (1973). *Science for the People.* London: Allen and Unwin.
LAYTON, D. (1991). Science education and praxis: The relationship of school science to practical action. *Studies in Science Education, 19*, pp. 43–79.
MEYER, K. (1991) '*Children as experimenters: Elementary students' actions in an experimental context with magnets*'. An unpublished doctoral dissertation, University of British Columbia, Vancouver, B.C.
MILLAR, R. (1987). Towards a role for experimentation in the science teaching laboratory. *Studies in Science Education, 14*, pp. 109–18.
MILLAR, R. (1989). What is scientific method and can it be taught? In WELLINGTON, M. (Ed.). *Skills and Processes in Science Education.* London: Routledge, pp. 47–62.
MILLAR, R. (1991). A means to an end: The role of processes in science education. In WOOLNOUGH, B. (Ed.). *Practical Science.* Milton Keynes: Open University Press, pp. 43–52.
MILLAR, R. and DRIVER, R. (1987). Beyond processes. *Studies in Science Education, 14*, pp. 33–62.
MILLAR, R. and WYNNE, B. (1988). Public understanding of science: From contents to processes. *International Journal of Science Education, 10*(4), pp. 388–98.
PERKINS, D. and SALOMON, G. (1989). Are cognitive skills context bound? *Educational Researcher, 18*(1), pp. 16–25.

PIAGET, J. (1969). *The Child's Conception of Physical Causality*. Totowa, NJ: Littlefield, Adams & Co. (first published in English by Routledge and Kegan Paul, Ltd., London in 1930).

ROGOFF, B. and LAVE, J. (Eds.). (1984). *Everyday Cognition: Its Development in Social Context*. Cambridge, MA: Harvard University Press.

SCHWAB, J. (1962). The teaching of science as enquiry. In SCHWAB, J. and BRANDWEIN, P. (Eds.). *The Teaching of Science*. Cambridge, MA: Harvard University Press, pp. 1–103.

SOLOMON, J. (1988). Learning through experiment. *Studies in Science Education*, *15*, pp. 103–8.

WHITE, R. (1988). *Learning Science*. Oxford: Blackwell.

WIGGINS, G. (1989). A true test: Toward more authentic and equitable assessment. *Phi Delta Kappan*, *70*, pp. 703–13.

WOOLNOUGH, B. (1991). Practical science as a holistic activity. In WOOLNOUGH, B. (Ed.). *Practical Science*. Milton Keynes: Open University Press, pp. 181–8.

WRAY, J. (Ed.). (1987). *Science in Process*. London: Heinemann.

Preamble to Chapters 7, 8 and 9 Approaches to Teaching Primary School Science

Maureen Duke, Wendy Jobling, Telsa Rudd and Kate Brass

Chapters 7, 8 and 9 describe the ways in which three primary school teachers, in three different schools, approached teaching the same science topic, 'soil'. In Victoria, science and other subjects are taught by generalist teachers who are not subject specialists. There is no specific science syllabus that schools are required to adhere to. Also, there are no state wide exams in science, or any other curriculum area, at primary school level.

The Victorian Ministry of Education has produced Curriculum Framework documents, including the Science Framework P-10 (Malcolm, 1987) to provide guidance for curriculum development at primary and junior secondary level but these documents are not prescriptive. Rather, they attempt to offer 'a clear direction for the curriculum, drawing on good school practices and current educational research'. The place of science in a particular primary school's curriculum is determined in a general sense by the school's management, but the amount and kind of science teaching and learning that students experience is largely determined by the interests, beliefs and practices of the teachers in the school.

From our experiences it is clear that teaching science in this context allows teachers, who are interested in science and feel confident about teaching it, to develop programs which are designed to stimulate students' interest in science and develop their knowledge and skills. However most primary school science teachers are not science specialists, and often do not recognize the value of science; even those who do may not have confidence in their ability to teach it. We believe that reflecting on and developing our views of science and science teaching has allowed us to become more confident as teachers. The lack of confidence that primary teachers experience is not only due to a lack of specific science training, although this is very significant, it is exacerbated by a view of teachers and teaching which casts the teacher as an authority who is possessed of a complete body of facts that has to be passed on unchanged to the students.

This is reinforced in the case of science as many primary teachers' view of science is based on memories of secondary school where science was a series of formalized recipe style procedures which seemed to be concerned with obtaining 'expected' or 'correct' answers. A spirit of inquiry was often not hinted at, much less aspired to. If you view science in this way, and regard teaching as a teacher-centred process where quantities of facts have to be passed onto students, then any perceived lack of knowledge is a daunting obstacle and engaging in teaching activities which might expose it can be very threatening.

We believe that if primary school teachers are encouraged to reflect on science and try out some science activities with their students they will realize that science can be far more interesting and rewarding, both for them and their students, than their own experiences as science students would suggest. To become confident they need to realize that they do not have to know everything, they must learn to say, 'I don't know but let's find out together'. This also has benefits for the students; it gives them confidence. The realization that learning does not stop, that everyone is learning all the time, even adults, helps to take the fear out of learning. Building the students' confidence so that they can become responsible for their own science learning is critical if they are not to be put off science from the beginning.

For this reason we try to ensure that our science lessons are student centred. This does not mean that 'any old thing will do' and the teacher's role is somehow diminished. For example, in the lessons described here our role in facilitating learning, questioning, sharing ideas, helping students to formulate their ideas and develop their skills was vital. As a teacher you can not only make use of the students' ideas, interests and experiences but can also expand the students' knowledge by leading them into areas outside their experience.

All of the teachers involved in this chapter share a constructivist/children's science view of science learning. As the lessons discussed here indicate, this shared view underpins many aspects of our teaching, while still allowing each to develop their own teaching style, emphasize aspects of teaching and learning that they consider important, cater for the needs of their students, and fit in with the ethos of the school.

We decided on 'soil' as a theme because, though some might see soil as too ordinary and therefore not having much potential, it is found everywhere and the students love it. Making use of readily available material is vital to primary school science lessons as there are no specialist science rooms, little or no equipment and no technical assistance is available. The students feel comfortable with soil and are interested in it, so you can build on that interest, find out what their ideas are and use them. It provides a starting point from which you can extend the students' knowledge in a variety of directions and develop valuable science skills.

Reference

Malcolm, C. (1987). The Science Framework P-10: Science for every student. Melbourne, Australia: Ministry of Education, Victorian Government.

Chapter 7

Primary Science in an Integrated Curriculum

Kate Brass and Maureen Duke

Rationale

In a primary school where an integrated approach to curriculum has been adopted, the subject of science does not exist alone. Subjects in the curriculum are not limited by boundaries; rather it is the interdependency of subjects which is utilized in the learning process. This integrated approach in this school is based on a number of rationale statements. These are, that children learn when:

- the environment is relevant and meaningful;
- the children are actively involved;
- they are able to negotiate their own goals and become responsible for their own learning;
- they are working in a cooperative framework;
- they experience a safe learning environment which encourages risk-taking;
- they see the teacher as a fellow learner or facilitator.

Staff at our school see learning as a continuum and are committed to the philosophy and principles of composite grades and mixed ability groupings. We have adopted an integrated skill-based approach to teaching and learning and plan topics on a whole school basis.

Focusing on Science

As outlined in the introduction to this chapter many of our primary school teachers come to the classroom with little or no background in science at either a secondary or tertiary level. As is often the case in the primary school we had tended to neglect or avoid science because of lack of teacher confidence and/or competence; yet all teachers were concerned and wanted to improve the situation. Our desire to 'do more in science' resulted in the running of an intensive 'Science Day' each term. These involved the children moving around in multi-age groupings to a number of different workshop activities which had been organized around a particular theme, for example, color, food, light and water. Teachers were able to support each other in developing activities which highlighted particular content. Staff still felt that four days per year limited the children, so those teachers who had attended science in-services returned to staff meetings to offer alternatives for the teaching of science. It was decided that the staff would meet for a planning day in December 1991, to plan topics for an integrated approach to curriculum.

We decided that we would choose a topic per term, one of which would have a science base. It would be in this term that the major science skills and understanding of science would be developed. This, in fact, means that a particular topic is common to the whole school. The science topic which was chosen, 'Our Place in Space', seemed to offer a wide scope for teachers to plan activities to develop specific science skills and understandings during the term. A focus question for the term was formulated, 'How can we take responsibility for the planet on which we live?'

A number of contributing questions to assist the answering of this focus question were similarly formulated:

Where is our planet in space?
What is unique about our planet?
What is wrong with our planet?
Who and what are endangering our planet?
What actions can we take to preserve/conserve our planet?

The contributing question, 'Where is our planet in space?' had been explored for a week, and the next question, 'What is unique about our planet?' led the children to the study of soil and its uniqueness on earth. The next two questions, 'What is wrong with our planet?' and 'Who and what are endangering our planet?' followed on from the lessons based on soil which are discussed here.

Encouraging Risk Taking

My class comprises thirty-one children in grades 2, 3 and 4 (aging from 6 to 9 years). When I observed the children at work early in the year in problem-solving tasks and decision-making situations, it became clear to me that the majority of them were not risk takers; not willing to guess or 'have a go' for fear of being wrong. Unless they were able to make guesses they would not be able to hypothesize. They were much happier to be spoon-fed information with the certainty of being right, to accept information without questioning. Because of this, I decided that the children needed to become 'guessers' willing to try without feeling failures and therefore become active in classroom science. By becoming active rather than passive learners, they would become responsible for their learning and 'own' what they discovered. The skills that I aimed to develop during the term were to:

(a) formulate an hypothesis;
(b) design an experiment to test the hypothesis;
(c) note changes which occur, and to record their findings in some way;
(d) observe using the senses, measuring, estimating, classifying;
(e) interpret the results and to draw conclusions;
(f) use a variety of research techniques in their investigations.

These science skills became the checklist for me to assess the children as they developed during the term. These were recorded in a book and ticked off as I saw that the child had mastered that particular skill. The science knowledge or content which I hoped to develop through an integrated approach to science was similarly noted down, as were the values and attitudes. Before drawing up any lists, it was important to build on what the children already knew, so there was a need for pre-assessment. Their skills had been identified. Their attitudes and values were indicated on a Likert Scale questionnaire, their knowledge was then assessed.

Pre-assessment: Exposing What the Children Know

To gain information, which I could use to direct the students to activities which would extend their knowledge, I conducted a number of pre-assessment activities.

First I conducted a whole class brainstorm about 'What is science' on the blackboard. The results are duplicated below (figure 7.1). Some of the children had very little idea, while others had no idea at all! The science definition from one child was 'it's when you be quiet' (suggesting silence, I suspect!); only a select few were responsive to this brainstorm. I prompted further discussion by asking 'What about scientists?' The children were a little more responsive as shown below. I did this because I think it is important for the children to realize that science is something that they can do, that they can be involved in. Like some primary teachers, children can be turned off science before they start. In order to gauge the children's pre-existing ideas about 'soil' I asked them to draw a glass jar full of soil. The younger children had a problem with the word 'soil' as they did not even know the word, so a quick trip outside enabled this to be clarified, 'Oh, dirt!' was their response. They preferred the word 'dirt', but it was discovered that dirt conjured up an idea of dark, 'muddy' type dirt, rather than a variety of soil types. In the results, most children drew black soil with no spaces in it, some children (for example, Susan, see figure 7.2) did have spaces in their drawing but it was not clear if these were intentional, and, if so, what they represented. I wanted to see if their drawings of soil really indicated all they knew about the topic, so I decided to question them further on the soil they had drawn. I hoped to discover anything else they knew about soil types and soil consistency. This questioning not only revealed that they knew far more than the drawings had shown, but also the misconceptions they had, as shown by the following extract:

(T = teacher, C = child)
T: Would you find anything else in soil besides soil?
C: Animals, worms, bugs, insects, twigs, leaves.
T: How do they move in soil?
C: They eat the dirt/soil. They grab it and throw it back. They eat the soil and then move on.
T: Imagine you are an animal who lived in soil. Do you need to move?
C: Yes, because you need to eat.
T: How would you breathe?
C: I would breathe the air.
T: Is there any air in your soil?

Figure 7.1: Brainstorm on 'What is science'

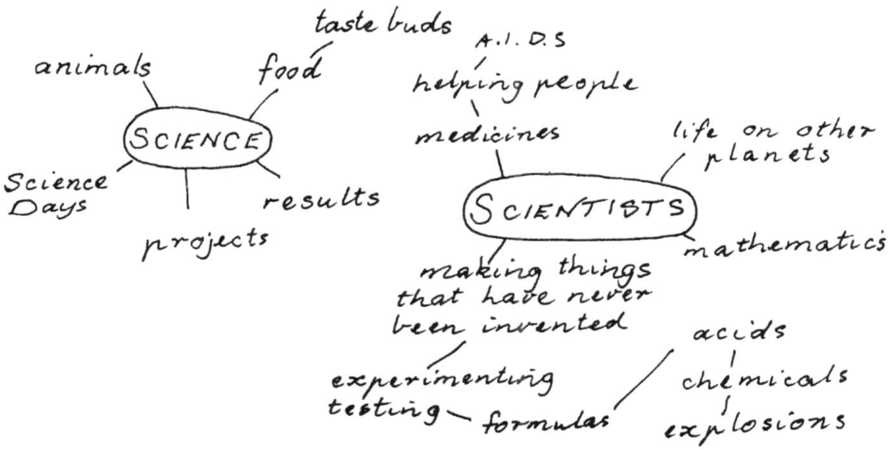

Figure 7.2: Children's drawings of a jar of soil

Paul **Susan**

C: Yes.
T: Where, is there any room?
C: I would stick my head up and breathe.
C: No, I would buy one of those things you use when you swim. I'd put it up high so it sticks out of the dirt so air can get in.
C: There is air at the top.

This discussion revealed that some of the students thought that animals living in the soil had to breathe by getting air from above the soil surface, rather than from air spaces in the soil.

After the children had expressed their knowledge through questioning and art work, a brainstorming session on 'soil' was conducted. I hoped that this

Figure 7.3: Brainstorm on 'Soil'

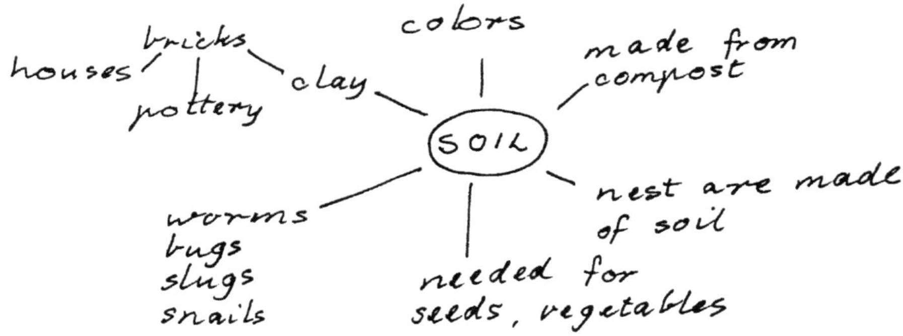

brainstorm would extend beyond the look of soil to its needs, purposes, etc. It was important that the children experience success at the brainstorming to provide positive experiences at risk taking. Without risk takers we can't develop responsible science learners (and owners of their learning) in our classrooms. This was done on the blackboard as a whole group so that the activity did not appear too threatening to some of the children. They were able to come up with a number of phrases/words/concepts. I questioned them as I had done in the previous activity. Their responses are summarized in figure 7.3.

Becoming Responsible Learners

From this questioning and sharing of ideas I asked the children what they would like to find out more about, or the questions they would like to have answered. I led a detailed discussion on each of the six questions they formulated with the purpose of directing them along the lines of guessing and forming hypotheses. The following are examples of parts of these discussions.

Student Question 1. Do seeds grow the same in different soils?

T: What do you mean by 'the same'?
C: Is it like a race?
T: What do you mean by a race?
C: Like when we all start together, but someone wins.
T: Why do you think the seeds might not all finish together?
C: The soil might be different.

My questioning continued:

T: How could you find out?
C: We could try it out. Someone could bring some seeds to school and we could plant them in some soil.
T: What would we have to do to get them to grow?
C: Give them water and light.

T: What do you think might happen, have a guess.
C: They won't grow the same because some soils are not healthy.
C: No, Some soils have got different chemicals.

Ten of the remaining children responded that there would be no difference in the growth patterns, yet were unable to give reasons for their guess. While these questions were being asked, some of the children merely sat, too scared to even guess. The suggestions from some of the other children for finding out answers to their questions were rather limiting: look it up in the dictionary/ask mum, etc. They were also not willing to have a guess at the result. When I suggested that I did not know either, they felt more willing to guess, knowing that they would not be told they were wrong. All too often teachers are unwilling to admit we don't know, or that we have made a mistake; all too often children believe that teachers know all the answers. The role of the teacher in this type of activity changes to one of facilitator, observing the interaction of the learners, stimulating them to think through ideas, asking questions to encourage hypothesizing. It is most important that we all become equals in the classroom, to allow the children to develop their skills at guessing, to feel free enough to 'have a go' without being ridiculed. This atmosphere must be developed in the classroom before any risk taking and any 'owned learning' can be achieved. It is vital that the children feel free enough to guess; without guessing we can't have hypothesizing! One child in particular had presented himself to me as a perfectionist. This view had been supported by his 'perfectionist' parents on their visit to the classroom during an open day. A perfectionist is rarely willing to have a go. It was evident that he would be one who would need a lot of encouragement to take chances and guess without worrying about the outcome.

Student Question 2: Can we change sandy soil into good soil?

C: You can water it.
T: If I went to the beach with a hose and watered the sand, do you think it would become healthy enough to grow trees?
C: No.
T: Do you know anyone who changes soil into good soil? What do they do?
C: They add compost to it.
T: How do they do it? Why?

The presence of the school compost bin was an ideal way to begin the discussion on composting and the changes which can take place. The children began to see that they could get information from one another as well as from books and adults as they were already actively involved in the composting 'business' at school. I went on to ask:

T: Does any one here have sandy soil at their place?

When the answer was yes from one child, and no from the child who lived next door, I asked, 'How come?' and the children began to discuss the idea of compost heaps. The discussion led into the properties of sandy soil and fertile soil. They were able to speak of sandy soil absorbing water when they water the garden.

Student Question 3: How long does it take for water to pass through soil and get to the bottom of the jar?

T: How could we test it?

C: We could get different types of soil and put them in a jar; put some water on the top and time how long it takes to get to the bottom of the jar. The jar would have to be made of glass.

Their question needed extending, as it was fine to find out how long the water took to seep through, but the 'why' question had to be posed by me. It is necessary for the teacher to encourage further investigation, to expand the thinking processes, to promote further experimentation.

Student Question 4: Do different animals live in different soil? If so, why?
The children were hesitant to even have a guess at this question, but they were skilled enough to make a number of suggestions when I prompted them by asking, 'Why don't we gather some samples from around the school; where could we collect them?'

A number of different places were suggested: outside the classroom window; from the plantation; near the compost heap; from the hill. They suggested that they could also collect soil from the homes of different people and see what animals there are in that soil. They suggested using an encyclopedia, the library, asking a parent or teacher to find out further information about the question.

Student Question 6: How do 'bugs' breathe, eat and move under soil? (By 'bugs' the children meant worms or insects.)
A number of the children had been involved in building a wormery, and so were able to share their knowledge with the others. When I questioned them further on verifying the information just heard, they came forth with a number of ideas to research the answer: use the library, ask someone, look up a number of books which they were familiar with on the class shelf. The students' explanations of how the animals breathed and moved included:

C: They eat the dirt.
C: They eat compost.
C: Other 'bugs' ('creepy crawlies').
C: They breathe through the cracks, eat the soil and move on.
C: They eat the soil and move on and they breathe with the spaces in the soil.
C: When they go into the soil they make a path out of the soil so the fresh air comes in.
C: Worms eat in as they go along and it comes out of the other side.
C: Through the holes.
C: They have holes in the soil and they eat the soil to get through.
C: They slither along.
C: They wiggle through the spaces.

Negotiating Goals and Becoming Secure Risk Takers

I invited the children to nominate which of the groups they would like to join to explore the answers to the other questions they formulated. It was great to see the children put themselves into a variety of multi-age groupings. It can be said that the children began to negotiate the areas they would like to study. It was left to them to decide which goals they wanted to achieve, and which areas of understanding or knowledge they wished to explore. I knew that there would be opportunities during the topic development and at the end of the topic for each of the groups to share their findings with the others and so become teachers and learners together. Active participation among group members and interaction between groups was encouraged.

They were encouraged to become active problem solvers — they were not being given the answers, but were encouraged to find out themselves. They were working in a cooperative classroom. In groups they felt non-threatened, secure risk takers as they joined with their friends to discover and learn together. Each time they worked in a group, they were given a different role to play, whether it be the reader, the recorder, the time keeper, the gofer (who *goes for* items needed). This was made possible through the children's prior experience with the cooperative learning framework. This enabled the children to become more responsible yet less alone. Had they each been left to find out their answers they might not have been able to develop this skill of risk-taking, guessing, having a go. I kept insisting that I didn't know the answers; this was needed for a number of the less confident children. The democratic framework upon which the school operates supported this development of responsibility, cooperation, mutual respect, self-discipline and social equality.

They decided that some of the questions could be grouped together. Each group was invited to discuss how they could find out the answers to their questions, and then to have a guess at the answers. Some drew, some wrote, but each group decided on the best way to tackle the solving of the problem they had set themselves. Some examples of their hypotheses and descriptions of their experiments follow.

Hypothesis A: That the darker the soil is the healthier it is, and the better the seeds will grow in it. (This followed discussion within the group on their knowledge of composts, and the plants they had noted growing around the school/at home in different areas).

Experiment A: The children gathered four containers of soil from various places, labelled them, planted seeds in them, gave each the same amount of water and placed them all in the sunshine. The children recorded the race between the seeds as they tested their hypothesis.

Hypothesis B: That different animals live in different soil because of camouflage, food supplies, ability to move. (This followed examination of the soil samples collected in the school yard, and other samples brought to school by different children. Not only did they find that the color and texture differed in the samples, so did the animals!)

Experiment B: The children brought a variety of soils from home and examined them for animals. They classified, compared and contrasted the animals in each soil type, this group divided into smaller groups as they researched the varieties of animals discovered the food they eat, the way they move, their color.

Figure 7.4: Students' second brainstorm on 'science'

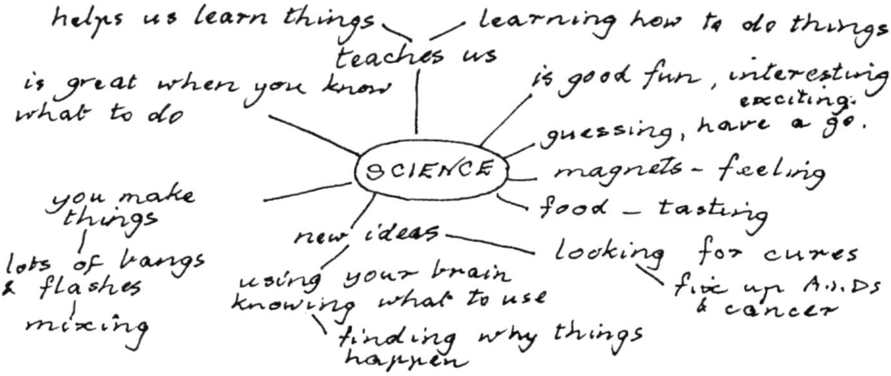

Sharing of Findings

The records were studied, the results were interpreted, and the findings of each experiment were presented by each of the groups to the whole class group so that all learners took on the responsibility of also being teachers for the class. In the past the teacher had all the answers, and the children were treated like empty vessels. The great benefit to this style of teaching and learning is that it moves the responsibility to the learner and shows that questions can be answered in more ways than one. Throughout these sessions I kept check on the knowledge and skill development of each of the children. They were encouraged and given regular opportunities to become self-evaluators during the unit development, by being continually challenged by each other and by the teacher; it became a game to continually question each other. Further assessment took place following the unit.

Post Assessment: Showing Knowledge and Skill Development

The children were again invited to take part in a brainstorm/concept map session on science (see figure 7.4). The children's second brainstorm was compared with the pre-assessment tool to gauge the change in knowledge development. There seemed to be a wider understanding of the place of science in their world. They did not have to be prompted as they had been in the earlier assessment (by brainstorming 'science' and then 'scientists'). It is important to note the two changes which took place by the end of the unit: the children began to understand that they can do science and that they can be involved in the study of science. When we begin to see ourselves as scientists we can do science and be involved in science; we can take chances, make guesses, hypothesize. Their responses reveal that they saw themselves as actively and more personally involved in science. Their earlier responses to the 'scientist' brainstorm were now included in their 'science' brainstorm showing the place of science moving from far-reaching

Figure 7.5: Students' individual brainstorm on 'soil'

scientific activities to their own involvement. Many more children responded to this brainstorm than had been the case earlier, showing the development of risk taking which it seems has taken place during the unit. It is also interesting to note the comments which indicate their attitudes and values to science, not merely knowledge.

The students were again asked to draw a glass jar of soil. These responses showed development from the previous perceptions of soil. No longer did they merely illustrate their jar of soil with solid black coloring. They showed significant development in their knowledge of soil, indicating a variety of living organisms in their soil and spaces in their soil.

The students were again asked to brainstorm 'soil'. On this occasion it was done individually. This gave them the opportunity to demonstrate their risk-taking skills as well as the knowledge gained during the unit. Some examples are shown in figure 7.5. The students' ability to answer the questions they had originally formulated themselves gave further evidence of the knowledge they had gained during the unit.

Skills Checklist

The skills checklist (referred to earlier) which was completed for each student during the topic development provided the record of skill development. It became evident that some of the children were more willing to have a go at guessing

Figure 7.6: Skills check-list

	Formulate hypothesis	Design experiment	Note changes & record	Observe using senses etc.	Interpret results & draw conclusions	Use variety of research tech's	
Anastasia			✓				
Andrea	✓	✓	✓	✓	✓	✓	
Billy				✓			
Stavros	✓	✓	✓	✓		✓	
Timothy	✓	✓		✓			

results, suggesting testing techniques, and recording results (see figure 7.6). Others of course still had a long way to go, but they were more comfortable with themselves and were at least able to take a more active part in the group work and saw themselves as worthwhile.

Activities in Other Curriculum Areas

Related activities in other areas of the curriculum gave the students a variety of ways to express themselves and enabled the teacher to further assess the skills, attitudes, values and knowledge developed during the term. For example:

maths:

* measure the liquid which passes through the variety of soil types
* measure the growth of each plant at regular intervals, and record these findings on a line graph
* discover patterns in nature
* measure set amounts of soil and compare the number of animals found

drama:

* pretend you are a particle of soil; get together with some friends and show how clay/sandy soil might be made
* imagine you are an animal in the soil; show how you move through/breathe in the soil
* think of all the things we use to move soil, and why we need to move it (spades, shovels, trowels, tractors, graders, trucks); mime the ways heavy soil/sandy soil might be lifted

process writing:

* imagine you have a treasure and you want to bury it; tell about your adventure as you search for the best place for

your secret; use one of the methods discussed in drama to move your soil

* a day in the life of a worm

art and craft: * free expression with clay led to the discovery of the properties of clay; the children were invited to make something of their choice

research: * the children were introduced to the variety of library resources we had available and were taught how to use the encyclopedia; where the science and gardening books were stored and how to use the catalogue

word study: * the spelling words for each week were taken from the word bank which was added to as each new word was met; the children learnt not only how to spell them but what they meant and how to use them. Because these words were introduced into the meaningful learning environment and had a purpose for the children, I found them being used in their process writing, language work etc.

language: * write a 'Who Am I?' and the answer, should be one of the things we have learnt about soil.

* describe the method I need to follow if I want to build a compost heap (you may like to write it in the form of a recipe, describing the ingredients, the method, the time for 'cooking').

Conclusion

In my role as a classroom teacher, I believe that I am a facilitator within the learning environment, and to that end, it is my responsibility to plan purposeful and meaningful experiences for the children. For that reason, the unit 'soil' could not be dealt with in isolation. While the knowledge and skills were science-based, I have endeavoured to show that the integrated curriculum has enabled the unit to stretch beyond subject boundaries, to give the children experiences which have had meaning and purpose and which have been developed over three weeks of working together. It is in the primary school where this flexibility is allowed; science is not a forty or fifty-minute period, but rather an experience which can be built on throughout the days and weeks of the unit development.

Chapter 8

Digging into Science — A Unit Developed for a Year 5 Class

Kate Brass and Wendy Jobling

This unit was taken with a year 5 class of twenty-four children in a government primary school located in a middle class area. Most of the children come from English speaking backgrounds. The school is coeducational with only one, mixed ability, class at most year levels. The school has quite extensive gardens consisting mainly of Australian trees and shrubs. Hence it provides a very good setting for science units such as this one developed around the topic of soil. Soil is an appropriate starting point as it allows a broad range of concepts and content areas to be explored. These include: soil composition, erosion, food chains linked to soil, and plant studies. Most children enjoy playing in dirt whether it is constructing roads for an imaginary city, looking for worms or watching ants carrying small insects to their nest. This unit of work makes use of this interest to extend children's horizons and develop their skills and knowledge.

Establishing Prior Knowledge

The children were asked to predict what makes up soil. They recorded their answers diagrammatically, as well as describing what they thought comprised soil. This step was taken in order to establish their prior knowledge. It is a useful technique, along with concept maps, for helping to gauge the best starting point for a unit of work. The teacher can ascertain what concepts and misconceptions children have with regard to the topic. He or she can then design activities to build on existing concepts and attempt to bring about changes where misconceptions are held. Also the children's diagrams and written predictions allow comparisons to be made with their actual samples and allow omissions to be seen and discussed without the teacher telling or 'correcting'. Ashley wrote, 'I think soil is made up of nutrients, dirt, worms, and stones.' Miles presented his prediction and follow up illustration as shown in figure 8.1. Most children decided that 'bits of rock' were an essential part of soil. Rotting leaves were also suggested. Most children, through the activities of their parents, seemed to be aware of compost.

Figure 8.1: Miles' prediction and illustration

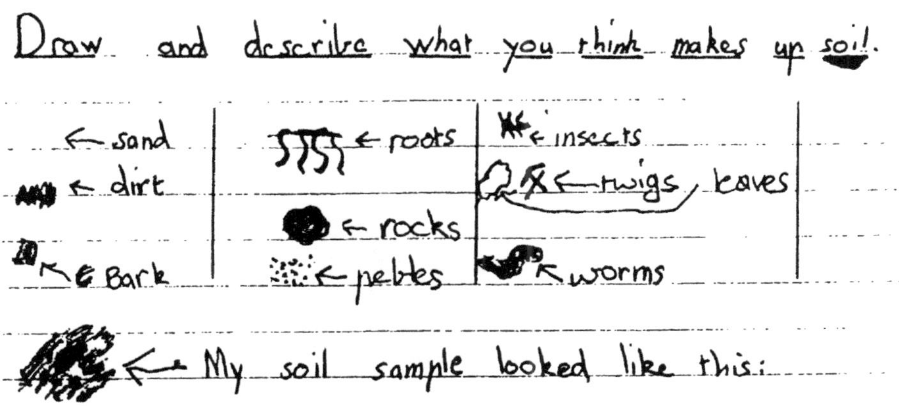

Observing and Recording

After recording their predictions the children then went out into the schoolground and collected small samples of soil. These tended to be from the garden beds. A hand lens was then used to examine the sample more closely. Children were asked to draw what they observed. The need for accurate recording became apparent to the children when Ashley and his partner found something wriggling in their sample. Naturally everyone had to have a look and ask, 'What is it?' This lead to a discussion about how we could find out. An accurate 'picture' would be needed, its size noted, and a written description of how it moved and where it was found. Reference materials from the library were considered as well as knowledgeable others such as parents and those in organizations such as the CSIRO (Commonwealth Scientific and Industrial Research Organization-Australia). These approaches were followed. At the time of writing the 'wriggler' has still not been identified.

Ashley's diagram is shown in figure 8.2. Although his written information on the worm (wriggler) provides useful information, his diagram of the soil lacks detail. It exemplifies the need to develop the skills of using written and diagrammatic methods to record findings. Soil samples were returned at the end of the session. The need for strict hygiene when handling materials such as this was also emphasized and acted upon.

The activities in this session confirmed some of the children's predictions as well as having them realize that small animals, such as the wriggler, and their remains could also be found in soil. As shown above, the task also served to lead children to ask their own questions based on their observations, thus giving them greater 'ownership' of their learning. Science activities need not depend on elaborate equipment or facilities. As well as soil, use can be made of available trees or shrubs within a school, even weeds growing through cracks in the asphalt.

Figure 8.2: Ashley's diagram of 'soil' and description of the 'worm'

Small Animals Found In and Near Soil

Following on from the previous session children collected small animals from the schoolground. Care was taken to note the location and to not harm any specimens. These animals were then drawn as accurately as possible. The need for inclusion of relevant details and not photographic representations was emphasized.

The following drawing of an ant (figure 8.3) is interesting as it shows that the child, Sally, has not observed the position of the legs on the body. This was discussed with the children. We looked closely at an ant and the parts of its body. A diagram of an ant colony was also examined. Sally's later drawing of a leaf insect (figure 8.4), completed as part of a project, showed the learning and skill development that took place with most children. It is interesting to note however, that some children still use stylized representations of an ant.

Children Posing Questions

After observing their animals children were then asked to draw up a set of questions to which they would like to know the answers. How these answers could be found, through both observation and through reference use, was also

Figure 8.3: Sally's initial drawing of an ant

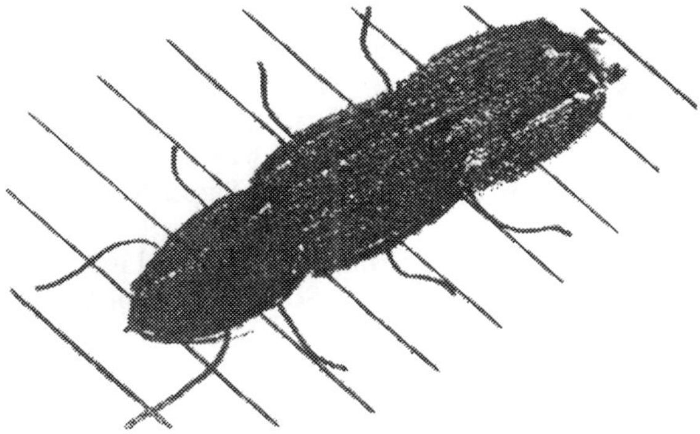

Figure 8.4: Sally's later drawing and description of a leaf insect

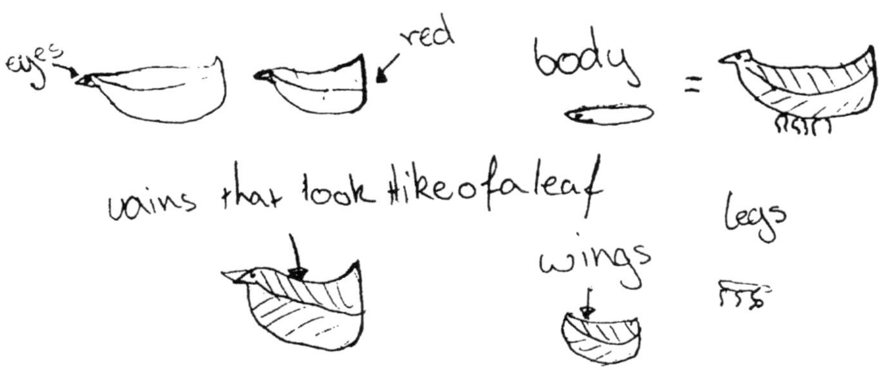

'A particular insect looked like a leaf. With six legs, yellow eyes and a red stripe in a semi circle going down from eye to eye. The colour of its body was light green the same as its wings. Near the bottom of its leg it's red when up the top it's green. The insect can fly and walk up and down the trees. In length it's 1½ cm and in width it's 1 cm.'

discussed. The teacher's role here is important as some children need guidance to enable them to ask questions of greater depth. Sally asked the complex question on the social organization of ants when she wrote that she was interested in finding out, 'If they have families like humans'. Her question, 'What they eat?' could with guidance be expanded to examine also how ants obtain and store food. Dan's similar question on slaters and Sarah's on spiders, could also be expanded in this way. The teacher could ask, 'How could you find out what they eat?' Here the child has the opportunity to suggest the method — perhaps observation, or a combination of observation and the use of reference materials.

The purpose of asking the children to pose their own questions was two-fold: firstly to promote their feeling of involvement in the task, and secondly to enable the investigation of questions that may not have been considered by the teacher. This teaching strategy has other benefits. It empowers children and allows the teacher to be seen as a learner and positive role model. The following examples of children's questions show several that would be unlikely to be asked by a teacher. Sally on ants, 'How old they live without being eaten or stamped on'; Dan on slaters, 'How its insides work'; Sarah on spiders, 'At what age do they leave their mothers?'

Below are some of Christina's set of questions and answers about a dandelion. (They are part of a project). Some of her answers show that she has made use of a reference book.

How is the seed carried away from the adult plant?
The seed is carried away by a person blowing the seeds, the wind blowing the seeds, or birds or animals carrying the seed on their feathers or hair.

How does the seed get onto the ground?
The seed is heavier than the parachute of hairs so it eventually falls to the ground.

How can we control dandelions?
We can control dandelions with weed pesticide spray or some better ways, nature (insects, rabbits and humans eat them), or we can just pull them out before the seeds get carried away.

Population Sampling

This task was given to give the children experience in using a 'scientific' technique and through this have them observe the link between soil, plants and animals. The method to be used was discussed with the children. Hoops were to be dropped backwards over their heads and then the plants and animals within the circle noted and sketched. (Naturally, some hoops will just happen to go rolling rapidly down the hill). This session resulted in very worthwhile interactions between children and teacher and among the children themselves. Children were again working in pairs which enables the teacher to discuss, at a more individual level, what has been found. One pair complained that there were very few animals within their hoop. I asked them why they thought this may be. They were not sure so we compared the hoop habitat, which was short grass on a play area, with the rather dense vegetation of the garden, and looked at the food and survival/ protection needs of small animals. This was done through asking questions such as: 'What are the differences between the garden and what is inside the hoop?'; 'What do small animals need to survive?' The purpose of this interaction was to get the children to think about the possible reasons for what they observed rather than giving them the answers. Later, when findings were compared and discussed, this pair shared their information with the rest of the class. This teaching strategy was used to enable children to make links between new and old knowledge. That is, children have to process the knowledge.

If children do not ask questions it may be necessary to initiate these through commenting on what various children found within their hoops. Questions such

Figure 8.5: Andrew's food chain beginning with maggots

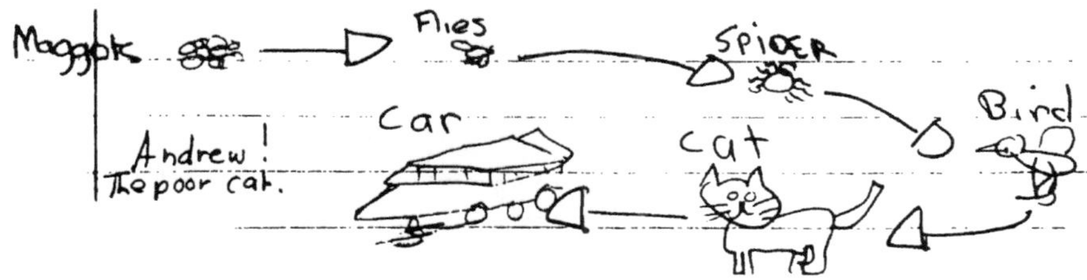

as, 'Why do you think very few animals were found?' can then be asked. Getting children to compare the garden area and the hoop area could also be useful. While outside each pair was given the opportunity to discuss their findings with the class. One pair found that their hoop was entirely filled with 'Cooch' grass. The manner in which this grows through sending out runners was discussed, which lead to the idea of plants being important in preventing soil erosion. (The area being studied was on a hillside which made this particularly relevant). Food webs, or chains, were discussed at this point and examples within the school ground suggested by the children. Two examples drawn sometime later show children's understanding of this concept (see figures 8.5 and 8.6). Cross-age tutoring sessions with year 2 children around the time of this activity were also linked to this unit. An 'Adopt a Tree' challenge was set for cross-age tutoring. Children from each grade level worked together over two sessions and a tree was selected by the team of year 2 and 5 children. On their sheet each group was to give a general description of the location of their tree. The tree was then sketched, a leaf and the bark texture drawn, or a rubbing taken of both. The purpose of the activity was to further develop observation and drawing skills and, particularly for the year 5s, to develop their ability to clearly communicate both verbally and through diagrams and written language.

During the following session these sheets were swapped around and, given both the written and pictorial information, each group had to locate the relevant tree. The importance of including relevant information was brought home to children through a heated discussion which took place during the second day. One group thought that they had located their tree. However, it was argued by one child that it couldn't be because the 'veins on the leaves in the drawing do not look like they do on the real tree'. The child drawing the original picture had accurately drawn everything else about the tree except for this element. The tree was a Pin Cushion Hakea with parallel veining. The argument was resolved through confirming with the artist in the original group that the correct tree had been found. This was a valuable peer group lesson for the child as it brought home the importance of looking for features that differentiate one leaf type from others.

Linking Soil and Plants Through Setting Up Experiments

Children were asked to design and make mini-greenhouses that could be placed on the window sill of the classroom (a link with technology studies). They were

Figure 8.6: Christina's dandelion food chain

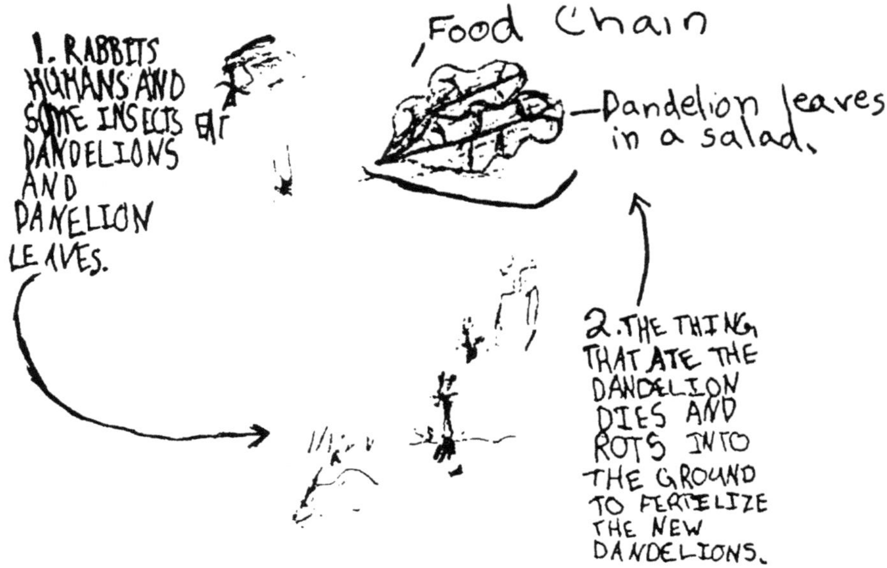

to consider the materials available and the size of the area. Working in pairs, or groups of three, meant that cooperation was essential. The design aspect of this task was linked to vocabulary and spelling work.

Children sometimes need to be introduced to cooperative learning activities in a more formal way, with each child in the group being assigned a specific role. As they become familiar with this method of working they can gradually be given the responsibility of assigning tasks themselves so each person has a worthwhile role. Initially a number of children in this class found it difficult to work cooperatively, but most can now work successfully with others.

The needs of plants were discussed with the children and they were then asked to design experiments to find out the best set of conditions for propagating the seeds of Australian native plants found within the schoolground. At this point it was necessary to have the children understand that a number of species need to be heated to allow seeds to germinate. This was done in the following manner. After the children had collected seeds from the playground the hardness of their outer coat was discussed. This led to a discussion of this being necessary because of bushfires. (Some evidence suggests that heat splits the seed's outer coat to allow germination. Another suggestion is that the heat of the bushfire sterilizes the environment). Ways of heating the seeds were then discussed. Miles was going to heat some on the barbecue while Sarah suggested the oven. Experiments were designed and set up by children looking at variables such as soil mix, heated and unheated seeds, the use or not of greenhouses. Two girls were looking at the inclusion of worms in one of their containers.

Assessment and Evaluation

Informal and formal methods of assessing children's concept and skills development were used. Firstly an assignment was set where children followed up several of the school activities described above. Children were able to demonstrate their understanding of concepts and skills learned. These included food webs, the drawing of accurate diagrams, writing useful descriptions and observations and using different methods of presenting data.

Several interesting ways of showing data were used. Miles used graph paper and plotted the position of plants and animals found in a square metre within his garden (figure 8.7a). Michelle presented her data in columns with a drawing, description of the actual size and the name of the plant/animal. Christina summarized the life cycle of a dandelion as shown in figure 8.7b. Another method of assessing children's knowledge and concept development was through the work presented by each child in the production of their group's book. The idea of the book was to communicate to others what was done and the results of activities. Children were asked to agree on who was to complete each section.

Assessment was also integrated with other areas of the curriculum to allow informal checking of children's learning. This included language activities where there was discussion of issues such as the role of introduced species in food chains.

Extension Activity

A science and technology studies activity leading on from this unit involved children designing and constructing nesting boxes for native birds and animals within the schoolground. Children watched the children's television program '*Behind the News*' (5 May 1992) where a researcher from Flinders University explained the need for these boxes due to the loss of hollows in trees for nesting. Our school recently had such branches lopped as they were considered to be a safety issue. Children designed the boxes after establishing the needs of the birds or animals they wished to help. This involved observation and the use of reference materials as well as contact with appropriate people and organizations.

The nest boxes were successfully completed and went on display at a State Park Visitors' Centre during August 1992 where they attracted considerable public interest. Permission has been granted by the School Council for the boxes to be placed in trees in the school ground. This is due to be completed in September with the assistance of parents. Children particularly enjoyed this activity because, as several explained, they were able to design and make something worthwhile.

Conclusion

This unit has allowed children to extend their knowledge and understanding of the role of soil in their environment. It has also provided them with the opportunity to follow their own interests while developing skills. These skills have included verbal, written and diagrammatic forms of recording and communicating.

Figure 8.7: Different approaches to presenting data (8.7a: Miles' work and 8.7b: Christina's work)

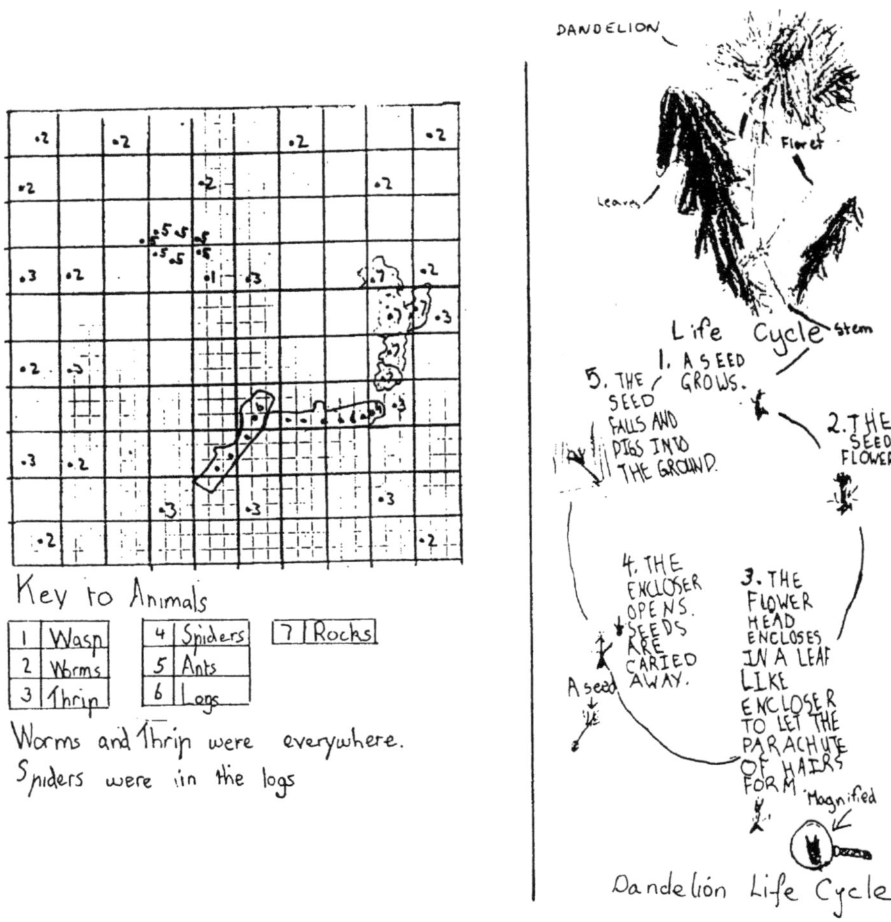

a: Miles' work

b: Christina's work

Chapter 9

Year 3: Research into Science

Kate Brass and Telsa Rudd

Investigation in science at every level in the primary school is the process of finding out as much as possible about a particular situation, using science as a tool for exploration. Investigation can be observed in primary school when children discover, for example, the many different ways to float an object. During my years as a primary school teacher I have observed that using investigational work in primary science will lead students to realize that science can be a challenge and that there can be a lot of enjoyment in thinking out scientific concepts for yourself. All children can think scientifically, to a greater or lesser extent, and even those children who find science difficult are very capable of thinking for themselves in their own way. One important aspect of children developing in this way is that the teacher's role in the classroom also develops. Changes in my role are described below.

Those children who seem to find the formal approach to teaching science difficult can excel when they are in an investigational and problem solving situation. They seem to take over the task with confidence and ease. The students participating in the lessons described here were aged between 8 and 9 years and were in a mixed ability class in a coeducational, independent primary school, situated in one of the suburbs of Melbourne.

Getting Started

At the start of the new school year I had a brainstorming session on the question 'What is science', so that I could discover the students' attitudes and feelings with regard to science. Doing this also allows the students to see that there are many skills needed to carry out scientific investigations and that these are skills they can develop in both science and other subject areas. The students recorded their ideas on large sheets of paper. The process of brain-storming is of value to the students because it allows them to develop their lateral thinking skills. It can also help them to develop other learning skills, such as listening, cooperative group work and asking questions.

Afterwards students were put into groups of three or four. Students at this age (8–9 year-olds) may need assistance when choosing the members of their group and I felt it was important for me to select the students for each group when introducing them to a new approach to teaching. I did this to help integrate

Figure 9.1: Brainstorming 'what is science?'

new members into the class; to encourage students to be aware of each other's strengths and weaknesses; and to encourage students to work with members of the class other than their close friends.

As the school year progresses and the students become more confident in themselves as independent learners and thinkers, they can be encouraged to select their own group members for other projects.

Each group added further ideas to the original brain-storming sheets. This gave me time to work with the individual groups encouraging further discussion amongst the students. Figure 9.1 shows the result of one group's brainstorming session.

Becoming Active Learners

Following on from this activity I gave the students a homework task requesting them to think of a question related to science to which they either didn't know the answer or knew only part of the answer. This activity gave me an insight into the scientific areas which the students wanted to investigate further. Their questions covered a wide range of topics including water, animals, plants, trees (environmental issues) and the planet earth, for example:

What stops the wind?
What are radio waves?
What are rocks made out of?
Why don't planets bump into each other?

How do clouds move?
What is gas?
Why does the earth go around the sun?
Why can some plants grow in sand and others can't?

At the commencement of the school year the staff had decided to cover six areas related to the science curriculum: matter, energy, life, time, change and relationships. I selected 'soil' as a topic which would develop the students' knowledge of the six science areas mentioned and at the same time take into consideration the areas of interest expressed by the students' in their questions on science. By using this approach the students were encouraged to participate in designing their own science programme, which I felt encouraged them to become active learners and helped them to develop the learning skills which would promote self-assessment (Rudd, 1992).

The learning skills which I wanted the students to develop throughout their science course included: Group Work – Cooperative group learning; Research skills, for example, developing library skills; Observation; Listening; Asking questions; Planning/Thinking.

It was also important to encourage the students to add more learning skills of their own to the above list as this assisted them to become independent learners. As the weeks progressed through the first term the students began to do this. They suggested: Being confident; Being independent; Making decisions; Sharing their findings with others; Taking care over the presentation of work.

The students kept a record of their progress on a chart which the groups marked off each day. They began to develop the idea that this list of skills was not only applicable to their science lessons but could be used in other areas of the curriculum, such as humanities, maths and language.

Choosing Topics for Research

Following on from this I held a second brainstorming session with the students to discover their views on 'How can we find out more about soil?' From this second brainstorming session the students and I selected seven topics relating to soil which the students wanted to investigate further by carrying out research projects. The seven research topics chosen were: Water and soil; Animals living in soil; A closer look at soil; How soil becomes soil; Growing seed in different soils; Tress (Rainforests); Setting up a wormery.

Class Lessons on Soil

Before the groups proceeded with their individual projects I carried out four activities on soil with the whole class. This was necessary so that I could help guide the individual groups and build up their confidence when embarking on their research projects.

Activity 1: What do you think is in soil? Students are asked to draw what they think they may find in soil (see figure 9.2).

Figure 9.2: Group drawing of what students think they may find in soil

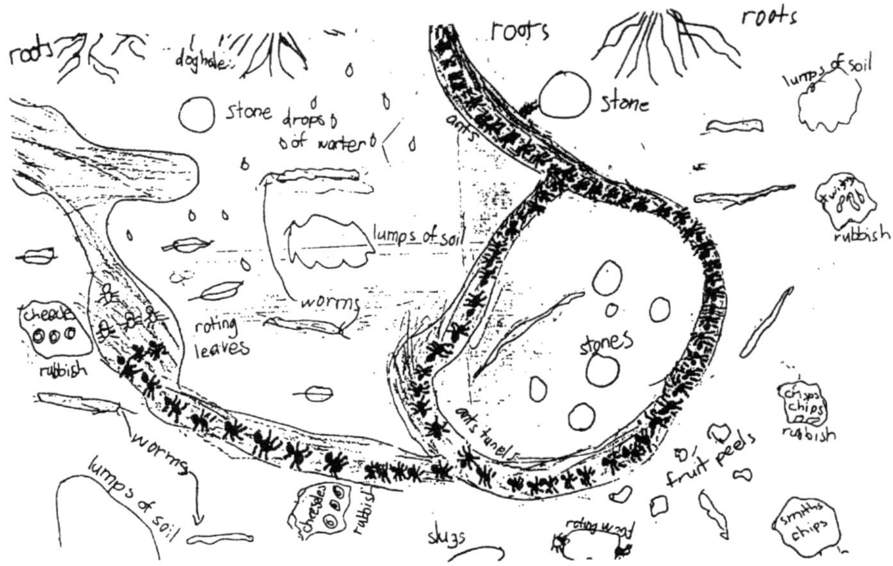

Activity 2: Students to bring a container of soil from their gardens, samples taken from their front and back gardens. Using hand magnifying glasses and microscopes (x5 & x10) students to examine their samples of soil. Students now draw a second picture of what they have discovered in their soil samples (see figure 9.3).

Activity 3: Teacher and students discuss their findings and compare their two drawings. The students were very absorbed in working with the microscopes and were amazed to discover that the particles of soil were made up of tiny pieces of rock, glass and sand with minute animals.

Activity 4: The students were asked to: Take a jar of water; add a handful of soil, stir well with a stick and observe what happens.

To summarize the usefulness of these four activities, I felt that the students had discovered a lot more about soil by investigating for themselves with the teacher's guidance than if I had used teacher centred activities. Also, they had developed skills by doing activities including learning to cooperate with individual group members, and, sharing their findings amongst the different groups. I believe the teacher's role at this stage involves responsibility for planning, explaining and guiding the students through the activities where the necessary skills are taught.

Developing Student Responsibility for Their Own Learning

As the students become more confident and competent learners, the teacher's role changes to that of a coach. The coach's role is to not only allow the students to carry out an activity but also to encourage the students to plan the tasks and

Figure 9.3: Students, drawing of what they found in their soil sample

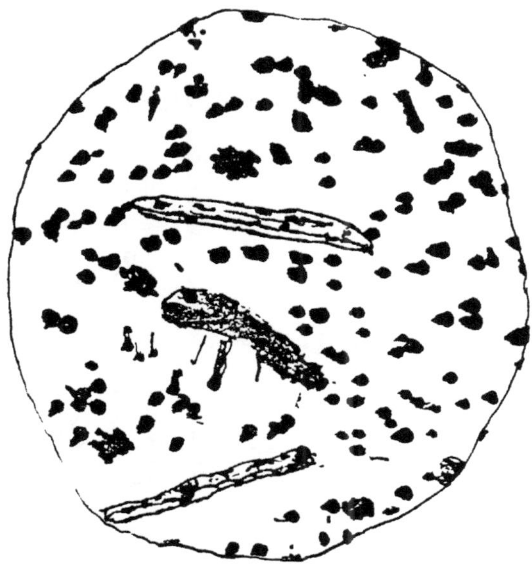

think about the skills needed. This stimulates students to think for themselves how the activity will develop. To encourage the students to be more responsible for their own learning I gradually increased the options for each project as the students progressed through their research by making suggestions, such as 'You may do. . .', 'When you have finished that piece of research on volcanoes, these are your choices. . .', 'Have you considered that approach to help you solve that problem?' Also by questioning the students about their methods and choices I felt that I could guide them towards other appropriate areas: 'What made you choose that tree activity?'; 'What did you learn from that activity?'; and 'Were you pleased with your results?'

I considered that giving the students these open-ended questions would develop their reflective thinking skills. They could express their opinions and ideas leading to greater creativity. In conducting this style of teaching I also wanted to encourage the students to design and plan out their own experiments, encourage the students to set their own goals, encourage the different groups to interrelate and share their research results and raise questions which could be answered by the other groups' results from their research.

I felt this approach would encourage them to become better self-assessors. This is supported by their comments expressed at the end of their self-evaluation sheets (figure 9.4). As students develop more responsibility for their own learning, the teacher's role now becomes a counsellor. The students are given tasks whereby they plan and think out for themselves the task and counsel with their teacher on their progress or problems arising out of the activity. The teacher can only reach this role when the students have confidence in and understanding of the skills needed to perform a task on their own.

Figure 9.4: Comments from students' self-evaluation sheets

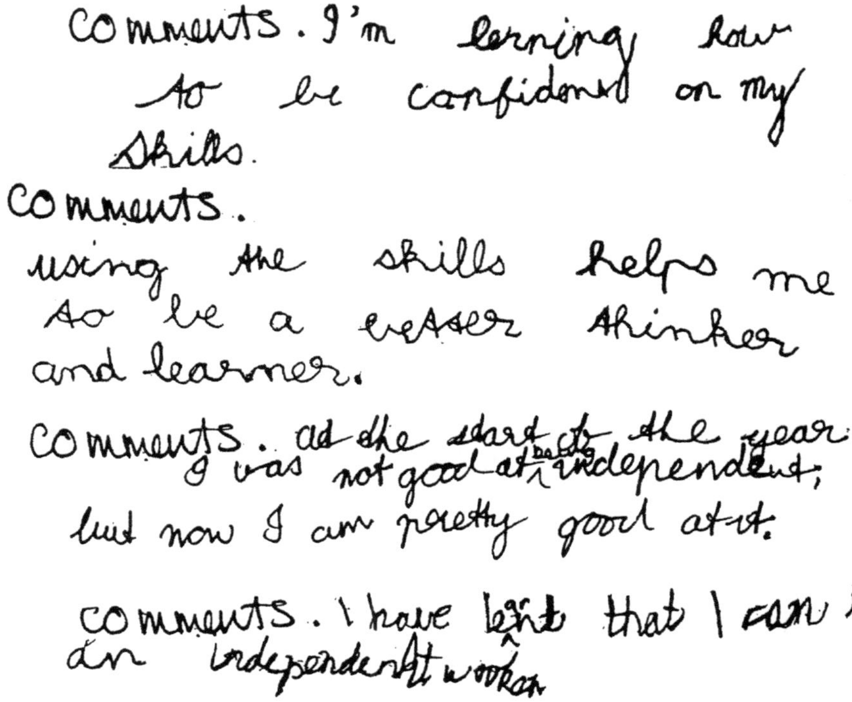
comments. I'm lerning how to be confidend on my skills.

comments. using the skills helps me to be a better thinker and learner.

comments. at the start of the year I was not good at being independent; but now I am pretty good at it.

comments. I have lernt that I can be an independent worker

Thinking Processes

Overall, to become independent thinkers and learners, I believe students need to develop thinking processes which can be summarized as follows:

Analysis: analyze; identify; explain; compare; distiguish; examine; contrast; investigate.
Synthesis: invent; improve; predict; plan; design; construct.

The final role of the teacher is that of a delegator. The teacher now allows the students to have complete control of carrying out an activity for themselves. The teacher in order to reach this role must have complete confidence in the students' ability to be independent learners and thinkers. This stage may not be reached until the fourth term or only when students are familiar with this approach to their learning.

In some cases I have found that only a certain number of students can reach this stage while other students remain at the second or third stages. Even when students have reached the fourth stage they may need to return to the other stages when they embark on a new activity or project. When the students become frustrated and argumentative, or when the activity becomes boring for the students, experience can help the teacher to know when to change her role and develop activities to meet the students' needs.

Ways to Assess Students' Learning

Working alongside the students enabled me to compile a checklist of learning skills of various kinds, such as: organizational skills, for example 'collect and return all equipment used in the activity' and 'plan my aims and goals'; self-assessment skills such as 'being aware of my strengths and overcoming my weaknesses', 'evaluating my work' and 'learning from the mistakes I have made'. Learning skills and cooperating skills would be 'sharing my results with other members of the group', 'helping other students when they need assistance' and 'working independently'. The check-list was used to form the statements on the students' self-assessment graphs as shown in figure 9.5. As well as the self-assessment check-list and graphs I have used the following to help me assess the students' progress throughout this and other units of work:

- Concept maps (figure 9.6);
- Careful observation, recorded on teacher and student assessment forms (figure 9.7);
- Questions and discussion;
- Performance of set activities;
- Ability to transfer learning;
- File of the group's work;
- Anecdotal records, interesting accounts collected by the teacher on the students and how they are learning.

For example, the following concept map (figure 9.6) was constructed by the whole class. It was mounted on the wall and added to by the students as their knowledge developed during the research projects.

Summary

Reviewing this method of teaching, I believe it allowed these young students to build up their skills as independent learners and thinkers and enabled them to take on the responsibility for their own learning. It allowed me to spend more time with individuals as they worked on their projects and I was able to learn alongside the students. I felt that the students learnt more for themselves and became more creative in their thinking when solving problems when I admitted that I could not answer all their questions. This method of teaching also encourages the students to see that other areas of the curriculum can be integrated with their science projects. As their research projects developed the students found that this led to more research. As the teacher I felt that something needed to be left unanswered as this encourages the students to develop their natural curiosity for wanting to discover more for themselves. I also noted that, as the term progressed, using this approach to teaching the students helped them to become more reflective in their thinking and better at self-assessing their learning skills.

Figure 9.5: Students' self-assessment graph

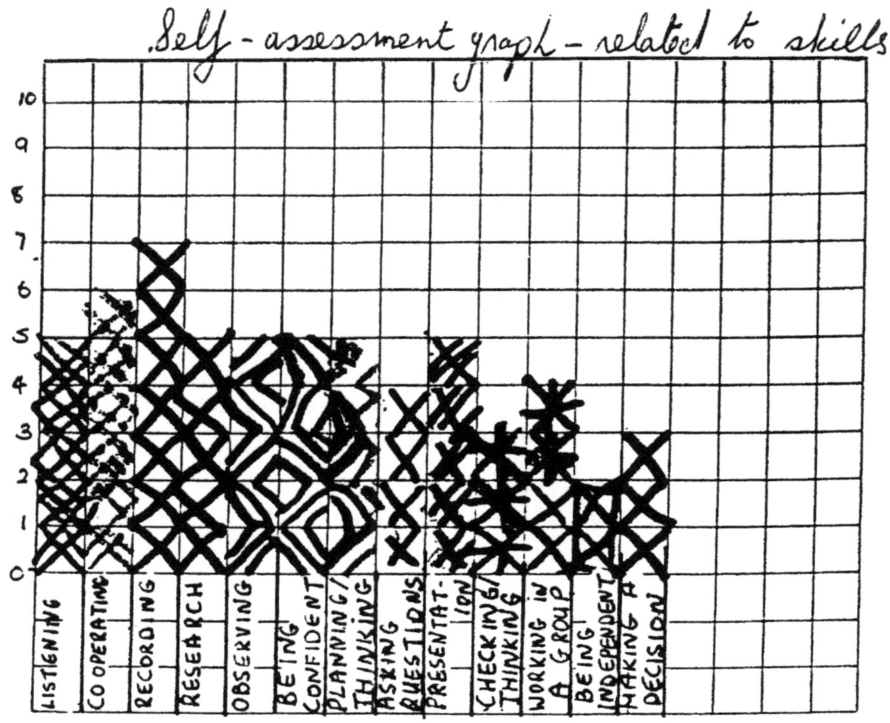

Much of the value in investigational work in science lies in allowing children to have more opportunity to be creative in science, in the same ways that they can be creative in the English language, maths and the crafts. Primary education should enable children to develop their understanding and their originality, as well as their skills. Teaching science in this way cannot start in the senior school, it must begin when the child enters the kindergarten, so that they can develop their investigational skills and strategies.

Investigational or problem-solving science is a complex subject and perhaps many teachers are apprehensive at the thought of having to consider it as part of their teaching. One lovely and reassuring comment can be found in De Bono (1972):

> Problem solving may seem to be rather a specialised part of thinking. But, if we change the name to 'dealing with a situation', 'overcoming an obstacle', 'bring about a desired effect', 'making something happen', then it can be seen that the thinking involved is very much the thinking that is involved in everyday life even though the actual problems may appear exotic. (p. 11)

Figure 9.6: Whole class concept map on soil

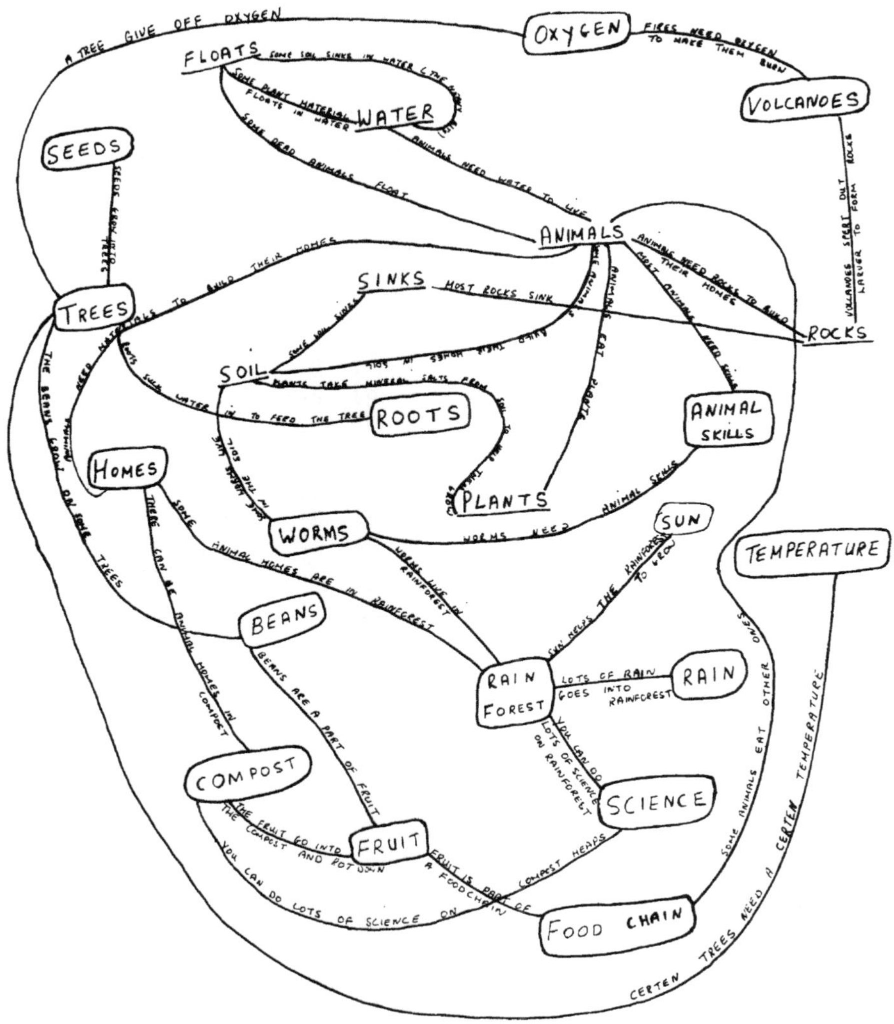

This passage sums up for me what every primary teacher is trying to do in their classrooms with the students they teach. So the fear of having to teach science seems insignificant; have a go and enjoy your science lessons.

References

De Bono, E. (1972). *Children Solve Problems*. London: Penguin.

Rudd, T.J. (1992). '*Exploring self-assessment with primary school students*'. Unpublished M.Ed. Studies project, Faculty of Education, Monash University, Clayton, Victoria, Australia.

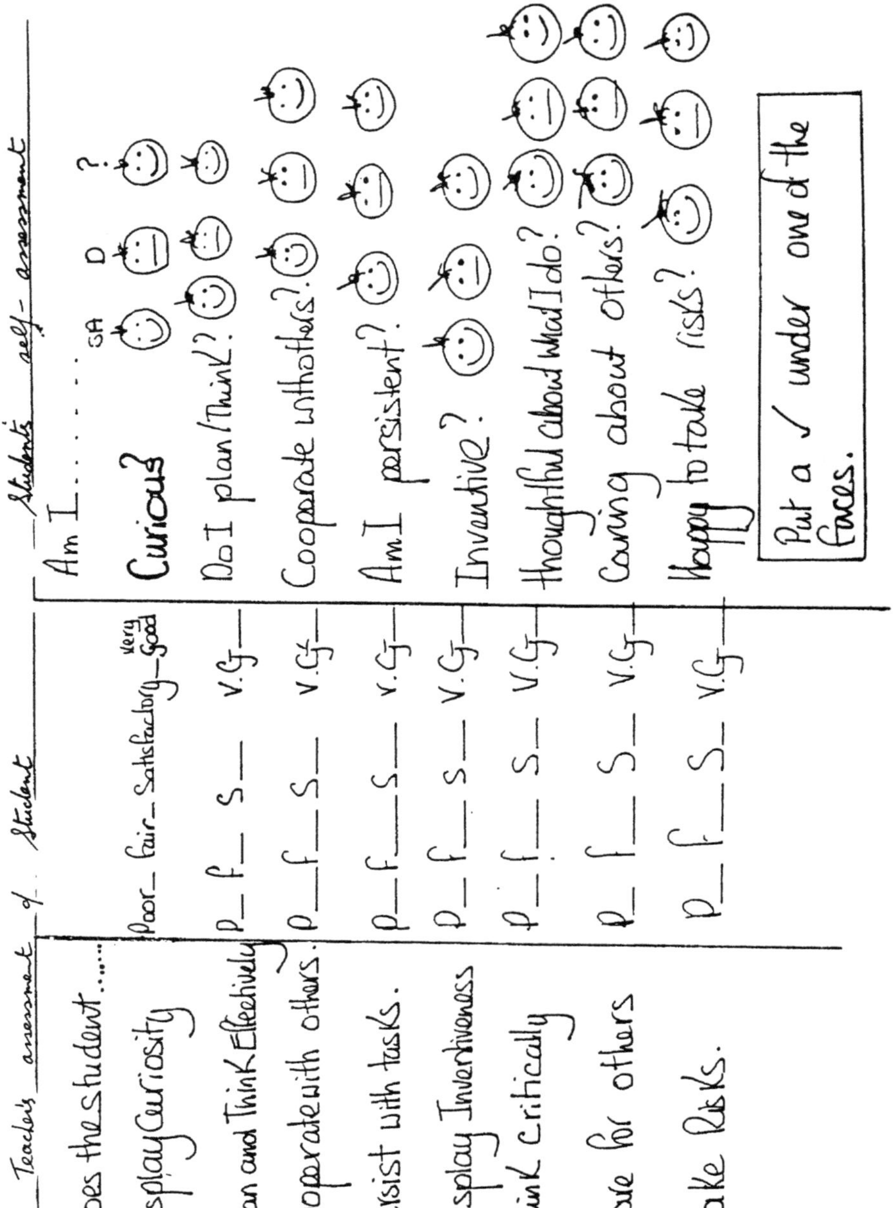

Figure 9.7: Teacher and student assessment forms

Chapter 10

The Importance of Specific Science Content in the Enhancement of Metacognition

Richard F. Gunstone

The essential thrust of this chapter is a tentative exploration of the nature of specific science content which can foster particular aspects of enhanced metacognitive behaviours among learners.

This thrust requires justification and explanation. Therefore the chapter begins with a brief discussion of the relationships between constructivism, metacognition and conceptual change, an elaboration of my meanings for metacognition, and some examples of the importance of metacognition (both 'positive' and 'negative') in conceptual change. I then explore some particular instances of the use of content in promoting more informed metacognitive approaches by learners, before attempting to analyze these content examples to suggest which features are important in the more informed metacognitive approaches.

Before turning to the first section of this sequence, I give two advance organizers of importance. First, the attempts to argue that content characteristics are important to developing metacognition are but an initial attempt. Our understanding of this issue is still inadequate. Indeed it is not at all clear that content/metacognitive development relationships can be taken any further than the instances given here, although I currently believe this will eventually be possible. Second, my small number of content/metacognition examples are derived from work with science graduates in a one-year pre-service high school science teacher education program. (The program leads to a Diploma in Education; hence these graduate pre-service students are termed DipEds from this point.) While these science graduates are, clearly, not high school learners, they do have one general characteristic of possible relevance to school contexts. Most of the Dip Eds have considerable propositional knowledge but inadequate broad conceptual structures relevant to this knowledge. Indeed, some can fairly be described as having grossly inadequate conceptual structures. Because of this characteristic I conject that the issues derived from the DipEd examples may usefully translate to senior high school situations where also the science learners already have propositional knowledge from previous science learning, but lack broad conceptual structures.

As you will see, the content in the DipEd examples is very typical high school content.

Constructivism, Conceptual Change and Metacognition

In a number of ways conceptual change, and how to promote this among learners, was a focus of this workshop which led to this volume. It is also the link which makes metacognition an utterly central issue to constructivist perspectives on learning.

For present purposes it is sufficient to take constructivism to mean that the learner constructs his/her own understanding from the totality of the experiences which he/she sees as relevant to the concept, belief, skill etc., being considered. For conceptual change, however, a more detailed description of my use of the term is necessary to show why metacognition is central.

Conceptual Change ('Replacement' and 'Addition')

The term conceptual change is commonly used to describe contexts in which learners hold existing ideas and beliefs which are in conflict with what is to be learned, and, hence, learners are involved in changing ideas and beliefs if they are to embrace what is to be learned. This usage is quite consistent with my use of the term here, although I will argue for a broadening of the nature of the ideas and beliefs to be considered.

'Conceptual change', as it has been used in the literature for some time, has usually meant the abandonment of one conception and the acceptance of another. One of the issues which was developed and clarified in the workshop was the expansion of this meaning for conceptual change (see chapter 1). Conceptual change may be replacement in the sense just described, or it may be the addition of a different conception. Addition here means an informed approach: an understanding of the value of the added conception in appropriate contexts. In this chapter I intend for conceptual change this expanded meaning of both replacement and addition.

Conceptual change, as we currently conceive it, involves the learner recognizing his/her existing ideas and beliefs, evaluating these ideas and beliefs (preferably in terms of what is to be learned and how this is to be learned), and then personally deciding whether or not to reconstruct these existing ideas and beliefs (Gunstone and Northfield, 1992). The reconstruction may lead to replacement or addition. This formulation of *recognize, evaluate, decide whether to reconstruct,* has clear links with the Posner, Strike, Hewson and Gertzog (1982) description of conceptual change as dissatisfaction with existing, then the new being intelligible, plausible and fruitful. The Posner *et al.* description has been a major influence on the thinking leading to the formulation above; in particular, by our work and reflection on the two elements of dissatisfaction and fruitfulness. Achievement of these is central to conceptual change, yet both can be very difficult to promote. What is significant here is that it is the individual for whom we seek conceptual change who must be dissatisfied, and who must see fruitfulness in the new conception. That is, the achievement of dissatisfaction and fruitfulness depends on the learner. This is the link with metacognition.

Metacognition and Conceptual Change

The description of conceptual change advanced above — recognize, evaluate, decide whether to reconstruct — clearly places the direct responsibility for conceptual change with the learner. This is inevitable, given a constructivist beginning. Of course it is obvious that there are major demands on the teacher in providing contexts where it is more likely that the learner will undertake these demanding tasks, an indirect teacher responsibility for conceptual change.

The essence of the importance of metacognition here is illustrated by negative cases where the context provided by the teacher can not have any impact on conceptual change because of existing ideas and beliefs about learning and teaching held by the learners. Consider, for example, a finding by De Jong and Gunstone (1988) from a study of attempts to promote conceptual change in senior high school physics students. Some students equated 'success in physics with two factors . . . high intelligence and good memory' (p. 23), and that these two factors were beyond their control. These were things learners either had or did not have. Thus, some of the students held ideas and beliefs about learning and teaching which precluded them from learning. Their ideas and beliefs led them to 'know' that they could not be successful learners of physics.

This is anything but an isolated example. Something similar is operating when learners say disheartening things to teachers such as 'we are having a discussion because you can't be bothered teaching' (that is, discussion is not teaching, I do not expect to learn anything because the teacher is not doing this to help me learn), or 'we are doing an experiment because it's Friday afternoon and you don't want to teach', or, 'if you understood this properly you'd be able to tell me so I understand'. Almost any science teacher who has seriously explored classroom teaching approaches derived from constructivist views of learning will have heard these statements. A number of examples are given in reports of the Project for Enhancing Effective Learning, PEEL (Baird and Mitchell, 1986; Baird and Northfield, 1992).

The significant point is that it is not only science related ideas and beliefs (see Carr *et al.*, chapter 11 in this volume) which impact on learning when learners bring these to science classrooms. Ideas and beliefs about learning, teaching and roles seen to be appropriate for learners and teachers all also impact on learning. In the above examples, learners are exhibiting ideas and beliefs about learning, teaching, and roles which are in conflict with those of the teachers and, hence, with each teacher's purposes.

My conception of metacognition is a rather multifaceted one. For the moment I describe it in the following terms: learners are appropriately metacognitive if they consciously undertake an informed and self-directed approach to recognizing, evaluating and deciding whether to reconstruct their existing ideas and beliefs. By informed, I mean recognize and evaluate, with an understanding of learning goals, of relevant uses of the knowledge/skills/strategies/structures to be learned, of the purposes of particular cognitive strategies appropriate to achieving these goals, of the processes of learning itself. Hence I argue that metacognition and conceptual change are totally intertwined. (This argument is elaborated in Gunstone and Baird, 1988.)

An Elaboration of the Usage of Metacognition

To illustrate my multifaceted conception of metacognition, I now give a number of different descriptions or examples. They are complementary, *not* alternatives.

(i) 'Metacognition refers to the knowledge, awareness and control of one's own learning' (Baird, 1990, p.184). Metacognitive knowledge includes knowledge of the nature and processes of learning, of what are effective learning strategies and when to use these, and of personal learning characteristics. Inadequate metacognitive knowledge restricts the extent to which personal awareness and control are possible. Metacognitive awareness includes perceptions of the purpose of the current teaching/learning activity, and of personal progress through the activity. Metacognitive control refers to the nature of the decisions made and actions taken by the learner during the activity.

(ii) Metacognitive knowledge, awareness and control are all learning outcomes as well as fundamental influences on the extent of achievement of more usual learning goals. That is, learners do not have some genetically determined levels of metacognition. Rather, the nature of personal metacognition derives from learning from experience (vicarious or managed) and hence, metacognition can be enhanced by appropriately designed learning experiences.

(iii) Often the learning which gives rise to a student's metacognitive ideas and beliefs has been unconscious learning, and the student then is unable to articulate his/her metacognitive views. However, it is clear from work such as the PEEL project (Baird and Mitchell, 1986; Baird and Northfield, 1992) that it is possible to adopt educational goals and to use teaching approaches which foster more conscious metacognition in students, and result in student learning about metacognition. This is true even for quite young children, for example, Rudd's (1992) work with her grade 3 class which resulted in these 8 and 9-year-olds producing complex and revealing concept maps about their own learning.

(iv) Examples have already been given in this chapter of metacognitive views which are at odds with learning goals involving conceptual change. One important generalization is illustrated by these examples: all students have metacognitive ideas and beliefs of some form. It is because of this that I refer to 'enhancing metacognition' when considering the development of more appropriate metacognitive views. (The conceptual change then implied by 'enhancing metacognition' is addition, rarely replacement).

(v) There can be tensions between metacognitive knowledge, awareness and control. Consider, for example, the extreme case of a course of study assessed solely via rote recall. Students who wish high grades and who have enhanced metacognitive knowledge and awareness will conclude that they should essentially relinquish metacognitive control. The text and class notes, in their literal form, represent exactly what must be learned for high grades. Put another way, any self-directed intellectual effort aimed at understanding is wasted when judged in terms of grades. (This example is also an important negative illustration of the point made earlier in this chapter about the major demands placed on teachers in generating contexts more likely to result in conceptual change. The context described in this example usually prevents conceptual change. The positive alternative is of course clear — change the assessment. Assessment consistent with learning goals is a fundamental feature of contexts that are structured to generate enhanced metacognition and conceptual change).

(vi) An appropriately metacognitive learner can be described as one who undertakes the tasks of monitoring, integrating and extending their own learning. Monitoring involves an informed awareness of both progress through task (which can be done via questions such as 'What am I meant to be doing?', 'Do I know what to do/write/look for?', 'What is the purpose of this task?', 'Have I done everything necessary?', etc.) and progress towards achievement of learning goals (for example. 'Does this make sense to me?', 'Could I explain this to someone else?'). Integrating and extending refer to the extent to which the learner links what is being learned with previous school learning, with existing personal ideas and beliefs, with applications/examples/ etc., in the 'outside' world, and with previous learning activities. Questions which can prompt this linking include 'What if — were changed?', 'Can I think of a situation where this does not seem to work?', 'If — is true, how is that — occurs?'

(vii) There are a number of very commonly occurring poor learning tendencies exhibited by learners (Baird, 1984 and 1986). These are superficial attention, impulsive attention, non-retrieval of relevant ideas and beliefs already held, staying stuck (i.e. one problem or error leads to no further progress), premature closure, ineffective 'unlearning', and lack of reflective thinking. These poor learning tendencies represent inadequate metacognition and are major barriers to learning.

(viii) Correspondingly, there are good learning behaviours which illustrate more appropriate metacognition in classrooms. These are many, and include: seeks clarification of specific points, including specific aspects when 'stuck' on a task; before commencing a substantial task plans a general strategy; checks work against instructions, then rectifies errors or omissions; is prepared to raise errors or omissions by the teacher; does not see the teacher as the sole source and controller of knowledge; takes relevant risks in order to increase understanding; considers the consistency of their explanations in different situations; etc. As already asserted, these good learning behaviours can be fostered by teaching.

(ix) There are complex and subtle links between metacognition and the importance to the learning of science of both the nature of science and students' epistemological conceptions (see again Carr, chapter 11 in this volume). This complexity is illustrated by some of the teaching examples given later in this chapter where the purpose of the example might well have been to explore aspects of the nature of science rather than aspects of individual learning.

I emphasize that the preceding nine descriptions/examples are complementary. These are not alternatives. This multifaceted view is also consistent with the way metacognition is discussed by Wittrock (chapter 3 in this volume).

Two Specific Examples of the Intertwining of Metacognitive and Conceptual Change

Conceptual change requires that learners are able to undertake, in an informed and independent manner, the recognition, evaluation, and decision whether to reconstruct existing ideas and beliefs. One obvious consequence of the lengthy descriptions of metacognition above is that the development of more appropriate metacognition should itself be seen as conceptual change — change of ideas and beliefs about learning/teaching/appropriate roles/nature of science.

The intertwined nature of metacognition and conceptual change is illustrated via brief references to two previous studies of conceptual change in physics.

In the first of these studies (Champagne, Gunstone and Klopfer, 1985), conceptual change was achieved with a group of DipEd students, in the area of mechanics. The beginnings of conceptual change in physics were marked by clear indications of greater metacognitive awareness by the DipEds., for example, 'Some people fight hard not to change preconceived ideas', '. . . the effort to hold out when I'm wrong is very draining', 'it's worth the lengthy time involved because I can have time to gradually understand the issue at point' (*ibid.*, p. 176). That is, as the DipEds became more informed about their own learning and the nature and purpose of the teaching, they became more able to recognize, evaluate, and decide whether to reconstruct existing ideas and beliefs relevant to physics.

The second study (Gunstone, Gray and Searle, 1992) represents an extension of this work with DipEds into intact high school classrooms. As well as exploring the value of the approaches to conceptual change in this more normal physics learning context, the role of metacognition was probed by withholding any attempts to explicitly develop students' metacognition. The conceptual change study was undertaken with grade 10 science students, and evidence of change in mechanics concepts for most students was clear. Some students then studied physics in grade 11 in another school, in a context where it was clear that their physics teachers were unaware of their grade 10 experiences. The teaching approach used in the grade 11 physics classes had no links with our study, and could fairly be described as very traditional. Even so, the cohort from the grade 10 study performed significantly higher on the grade 11 mechanics assessment than did the cohort of other grade 11 physics students. The more important findings here, however, came from the interviews conducted with the cohort from the experiment after the grade 11 mechanics assessment. These interviews explored student perceptions of the grade 10 experience in terms of recollections of the experience and other grade 10 science experiences, purposes of the experience, differences in approach in the experience as compared with the rest of grade 10 science, and the value to their grade 11 physics learning of the experience. Student perceptions of value varied from extremely helpful to negative impact on grade 11 learning. These variations were clearly dependent on both the extent to which the student recognized the purpose of grade 10 and the extent to which this purpose was accepted. Where the purpose was not accepted this arose from a rejection of the view of learning on which grade 10 was based (for example, '... although we don't know what we are learning we still get through all right', p. 185) and/or a rejection of the nature of physics implied by the qualitative approach in grade 10. In the extreme case of this last point, three students at one of the two sites at which the study was conducted denied any conceptual link between the grade 10 experience and grade 11 physics.

For me, this second example clearly illustrates the interdependence of metacognitive/conceptual change by showing that for some student metacognitive insights allowed them to see and accept grade 10 purposes. Hence, they were able to build on an existing conceptual understanding in their approach to grade 11 physics. Those students who saw grade 10 as a disadvantage illustrate two relevant points which have been constant threads in the work of John Baird (for example, Baird 1986 and 1990; Baird and Mitchell, 1986). Poor performance may often be due to poor learning approaches rather than an inability to learn. Learners are often unaware of their own learning tendencies and their own level of understanding.

Examples of Specific Content and the Enhancement of Metacognition

In this tentative exploration of the features of particular content contexts for enhancing metacognition a small number of examples is considered. For each example, the metacognitive purpose is described, the content and pedagogy briefly indicated, and some sense of the metacognitive outcomes is given. Then the content is analyzed in terms of the features which appear to be of significance to the metacognitive outcomes. The examples have been chosen because they represent successful practice, that is, because there is evidence for suggesting that their use has contributed to enhanced metacognition for at least some of the participating learners. The examples are from the Monash DipEd program. The broader context of this program has been described on several occasions, most recently by Gunstone and Northfield (1992).

Example 1 — Normal Reaction

Metacognitive purpose: This is the first specific attempt to achieve metacognitive aims in the science DipEd program. As a consequence the metacognitive purpose is somewhat more general than for examples 2 and 3. The broad purpose is to have DipEds recognize their ideas/beliefs about learning/teaching/roles and, since this recognition is in the context of a successful alternative learning experience, to begin evaluation of these personal existing ideas/beliefs.

Content and pedagogy: An example of the teaching sequence is given in figure 10.1, an example in two senses, namely, some of the detail shown in figure 10.1 (for example, 3 (e): gravity pulls the same on all things) is more prevalent

Figure 10.1: Outline of normal reaction sequences

1 Establish as existing knowledge: force is push or pull, direction matters.

2 Book on table: Force(s) on book? (Gives two groups: gravity only and gravity + table; latter group largely see the two forces as equal).

3 (A sequence to challenge the basis and conviction of those asserting two forces in step 2)

   (a) Student (small) sits on chair with feet raised—Forces? Equal?

   (b) Teacher (189 cm tall, over 85 kg) sits on chair—Forces? Equal?

   (c) Compare push up of chair in (a) and (b). (Result: commonly these equal, this not a problem as gravity pull same on all things)

   (d) Student, then teacher, stand on bathroom scales. (Result: commonly teacher heavier, gravity pull the same)

   (e) Gravitation related to particle model. (Result: commonly weight is gravitational pull, gravity pull different on different things)

   (f) So, how does the chair know exactly how hard to push up? (Left, for the moment, unresolved)

4 Question from 2. (Force(s) on book?) for one book on outstretched hand, several books on outstretched hand, one book suspended from spring, book on table.

5 Metre ruler used as model of table, book on ruler—Forces?

6 Analogy drawn with a long bow (including tactile responses)

7 Results in agreement that push up by table on book is believable and explicable if table does distort.

8 Show table distortion (mirror, slide projector to magnify).

9 Resolve chair issue.

10 Link with relevant experiences: weighing tables in chemistry, houses where stumps are missing, etc.

for some uses of the sequence than for others, and some strategies (for example, 4: forces on book(s) in a number of contexts) assume more prominence in some uses of the sequence. What is shown in figure 10.2 is the sequence as it happened with one group. The sequence is used with relatively small groups (sixteen-twenty) containing a mix of biology, chemistry and physics majors and is timetabled in the first or early second week of the science DipEd program. It is for most of these students their first experience of teaching with explicit conceptual change goals.

The sequence is strongly derivative from that described by Ministrell (1982). Hart (1987) describes a similar sequence. In our use with DipEds there are some variations and extensions. Some of these (3. in figure 10.1: consideration of

Figure 10.2: A selection of responses to the question 'What have I learned about my own learning?' written after the normal reaction sequence

1 'It (i.e. my learning) needs a lot more thinking about.'

2 'I can't understand anything about physics.'

3 'I may not know as much as I thought because when asked to explain to others, problems and self doubt occurred.'

4 'I have previously been satisfied with explanations that I haven't fully comprehended in tangible terms.'

5 'I can be confused easily. I would rather accept the facts than try to work out the reasoning behind the theory.'

6 'What I learned was influenced by what I already knew and believed.'

7 'I have always found it difficult to remember what I was taught but easy to remember what I learnt (that was from a Patrick White novel). Therefore I need to be challenged to grasp an idea myself rather than have it handed to me.'

8 'Something that seemed obvious I couldn't explain and this annoyed me. The limitations of my vocabulary (as a physics student) was considerably inhibitive.'

9 'If I ask questions I seem to understand better—and I'm not afraid to ask!'

10 'Even though I felt I contributed little to the discussion (which disappointed me a little—I felt my comments would have been of little use) the lesson did get me thinking. I formulated my own ideas as the lesson progressed.'

11 'I feel I have to challenge others' ideas (either outright or just to myself) in order (to learn).'

12 'I had misconceptions about ideas I thought to be fact.'

13 '. . . I can learn much more by explaining to someone who doesn't know . . .'

14 'I tend to see what I already believe.'

15 'It may require unlearning.'

variable reaction forces from a chair; gravity in the context of a particulate model of matter) are specific inclusions for the DipEds. The first (the chair) is intended to challenge those DipEds who assert that there are two forces on the book, but do so without any conceptual understanding (i.e. by asserting a rote learned $F = ma$). The second inclusion draws on the fact that we have not yet found a science graduate who does not embrace a particulate model of matter, hence this model is a universally accepted starting point for considering gravity forces on different objects.

The pedagogy used in the sequence is, with one exception, interpretive discussion (Barnes, 1976). The exception is 3 (e) in figure 10.1, the linking of gravitational pull with particulate model, which is done via teacher exposition and teacher-directed questioning.

Metacognitive outcomes: There is a cognitive issue here of major significance to the metacognitive outcomes. After this relatively short (two-hour) experience, most DipEds see that their understanding of the normal reaction force is substantially enhanced. This cognitive outcome, which relates to a feature of the content discussed in the next section, is most important in that it promotes introspection among the DipEds about their own learning and the learning of others. Such introspection is, as far as can be achieved, 'demanded' by the concluding tasks in the sequence. Each student writes answers to 'What have I learned about physics?', 'What have I learned about my own learning?', 'What have I learned about the learning of others?', 'What links are there between my learning or lack of learning and the teaching approach?', plus 'Any other comments'.

A selection of responses to 'what have I learned about my own learning?' is given in figure 10.2. Before describing something of responses to the other questions, and subsequent use of the responses, some comment is made about the responses in figure 10.2.

Some of these responses are likely to represent recognition of aspects of personal learning (for example, nos. 1, 3, 4, 5). Others are likely to represent at least the beginnings of an evaluation of aspects of personal learning (for example, nos. 6, 9, 11, 12, 14, 15). In some cases (nos. 2, 8) an affective issue has resulted in the experience not having value, cognitive or metacognitive, for these individuals. And some (for example, no. 7) are likely to express views already formed before the experience.

One other feature of the responses in figure 10.2 is that there are statements that would have been central to any set of lectures on constructivism and metacognition (for example, nos. 4, 9, 12, 14). Many responses to other questions also have this attribute. Examples about the learning of others include: '. . . I think we all gained *an* understanding', '. . . people may not realize the prejudices/biases they have', 'Others will see something in a totally different view to me', 'Just because I understand doesn't mean others can "know" what I am saying if I don't use their "language,"' 'It is hard to be rid of preconceived notions', 'Previous biases and so-called learning can hinder new learning instead of helping it'; for links with the teaching: 'Methods of teaching . . . must be relevant to the person's experience. Some people share things that others can't relate to', 'First the problem was posed and I knew what I didn't know'. In the extreme case of these examples of insights into learning, one student in 1992 wrote:

> (The teaching was an) example of teaching a fundamental aspect of physics through a very clear and step by step build up in confidence of beliefs. My only concern is that about a month from now — lacking any reinforcement of the concept — some people will forget how this explanation fitted together and will return to their original belief discounting this as a time when they were *led* into believing what (the teacher) wanted them to believe.

This is a clear account of the possible difficulties with long-term understanding after conceptual change teaching, which have been so well described by Gauld (1986).

The metacognitive outcomes of the experience are shown by these responses. These are built on by returning to the DipEds a copy of all these responses, and then using them at appropriate times during the rest of the program.

An analysis of the content in terms of metacognitive outcomes: The important aspect of the content used here is that it is possible for most students to make significant cognitive gains in a short time. This is crucial as it is only for this reason that many DipEds are prepared to invest the considerable intellectual effort required to seriously participate in the experience. In the language of Posner *et al.* (1982), fruitfulness is achieved in terms of the cognitive outcomes.

Some other relevant features of the content are obvious. DipEds have much relevant propositional knowledge (not always consistent with science) but poor understanding. Indeed they are often well aware of their poor understanding and may even see this as, somehow, inevitable - 'I can't understand physics'. Comment 2 in figure 10.2 indicates that this perception of inevitability can remain after the experience.

These features — relevant propositional knowledge and poor understanding — may be necessary but they are not in this case sufficient. The characteristic of the content which is most important to the rapid cognitive gain is, crudely, that the conceptual problems of the DipEds derive from the physics perspective on normal reaction rather than the DipEds' existing ideas and beliefs. The essence of the teaching sequence is to show the cause of the normal reaction force (distortion of surface), something rarely addressed in physics texts or in teaching. Given that the major conceptual barrier to learning is in the way the physics is presented, then this allows a situation where fruitfulness can be seen in a relatively short time. Without this short and positive time many DipEds would withdraw from active intellectual engagement with the sequence as, with their experience being their first of this nature, they did not see that cognitive fruitfulness would be as achievable beforehand.

Example 2 — Falling Bodies

Metacognitive purpose: In this case there are much more specific metacognitive purposes than was the case for the first example. The first is to point to the theory-dependent nature of each student's own observation. This is something of particular importance to intending science teachers. A second is then to have a base from which subsequently to move to some of the epistemological issues surrounding observation and science.

Content and pedagogy: The pedagogy used is Predict — Observe — Explain, or POE (White and Gunstone, 1992). Learners are shown a situation, asked to write predictions and reasons, some discussion of predictions and reasons is undertaken, observations made (and written), and any necessary reconciliation between prediction and observation undertaken. This can be with either large or small groups, and both of these modes have been used with DipEds in this case. The size of the group impacts on the extent and nature of discussion.

Two POEs are undertaken, both involving a shot put and rubber ball of very nearly the same diameter, with the mass of the shot put about thirty times the mass of the rubber ball. The first POE involves predicting the relative times of fall when dropped from about two metres, the second POE is identical except the balls are dropped from 9.6 metres.

Figure 10.3: Predictions and observations for two falling bodies given by twenty-four science graduates in DipEd

	OBSERVATION			
PREDICTION	SAME TIME	SHOT FIRST	?SAME OR SHOT FIRST	CANNOT TELL
SAME TIME	7	—	—	—
SHOT FIRST	2	10	3	2

Metacognitive outcomes: The first POE (dropping from two metres) is central to the specific purpose of showing the theory dependence of observation. Figure 10.3 illustrates the outcomes of prediction and observation with one group, a very typical set of data. These show clearly the impact of prediction on observation, a hard point for any DipEd to ignore. Reasons for predictions are also important in that these show the variety of 'theories' which influence observation. In the case of the data shown in figure 10.3, reasons given to support prediction of shot put first were shot has greater weight, air resistance affects shot more, have tried this and seen what happens. Reasons for equal times were gravity pull is equal on all things, have tried it and seen what happens, have been told this/taught this/ read this.

The second POE (dropped from 9.6 metres) is an important extension which allows the start of a consideration of observation in science. Although predictions vary between equal times and shot put first, the observation is absolutely clear to all - the shot arrives first. This then leads into a discussion of the role (or otherwise) of empiricism in formulating the generalization that all things fall with the same acceleration, and some consideration of the value and purpose of the generalisation. The myth of Galileo, the Leaning Tower of Pisa and the supposed dropping of objects, is also then addressed.

An analysis of the content in terms of metacognitive outcomes: There are two obvious characteristics of this content which make it appropriate for the central metacognitive purpose. First, the observation is contentious. It is not clear which, if either, of the two balls arrives first when dropped from the lower height. Second, this is a common phenomenon, and one for which DipEds have both considerable propositional knowledge (again not always in accord with science, or even always relevant) and links with everyday experience.

Example 3 — Ohm's Law

Metacognitive purpose: As for the second POE, this one's purpose is quite specific — to show the importance to cognitive learning from laboratory work of knowing and understanding the purpose of the laboratory exercise.

Content and pedagogy: This is one episode at the end of a sequence. The whole sequence, which involves biology and chemistry majors learning about DC circuits and relevant concepts from high school science programs, is intended to both develop conceptual understanding and to have DipEds experience, as learners, a variety of relevant teaching strategies. A description of the sequence is given in Gunstone (1991). That description is of an earlier usage where this metacognitive purpose was not central to the laboratory exercise.

This laboratory exercise involves measuring V and I across an incandescent globe for various values of V, and then graphing these data and drawing a conclusion. Prior to undertaking this experiment the sequence had focussed on developing an understanding of potential difference, current and resistance through the use of a number of analogies, discussion, POEs and various other tasks requiring active processing by the DipEds.

One aspect of the earlier consideration of these concepts involved a discussion of the effect on I of increasing V. This led to a statement of Ohm's Law, a piece of propositional knowledge that was familiar to participants in its symbolic form. Both in this discussion and in the printed notes the teacher asserted that Ohm's Law did not apply to all situations in which current flowed as a result of a potential difference.

Immediately prior to the DipEds commencing the laboratory exercise, the teacher asserted the purposes of the exercise to be two fold: to have the experience of wiring an actual circuit as a form of reinforcement of aspects of the sequence, and to explore whatever relationship there might be between V and I for the globe. After all had collected and graphed data and drawn their conclusions, each then wrote an answer to the teacher's question 'What was the purpose of doing this experiment?'

Metacognitive outcomes: These are derived from discussion of the cognitive outcomes (the graphs and conclusions drawn by the Dip Eds) and the answers to the metacognitive question about the purpose of the experiment. One representative example of a common product is shown in figure 10.4. The features of this example are the straight line (usually described by the DipEd as a 'line of best fit') drawn through a set of points which appear to fall on a smooth curve, the ignoring of what could be argued to be the most reliable data point (0,0), the conclusion, and the explanation for all of this in the perception of purpose. Approximately half of the group in the most recent (1992) use of the sequence concluded that V was proportional to I and/or that Ohm's Law had been verified, sometimes by claiming 'experimental error' or by 'allowing for errors in the equipment'. All but one of these Dip Eds explicitly asserted one of the purposes of the experiment to be the verification of Ohm's Law. (In a previous use of the sequence one DipEd achieved a straight line graph by adopting a non-linear scale on the horizontal axis).

The subsequent discussion of results focussed on the nature of graphs and conclusions, and the links with perceptions of purpose. When the teacher asked whether the purpose was stated beforehand, this was universally denied. At times, this discussion has become a little hostile as DipEds assert that it was reasonable for them to assume the purpose of the experiment to be the verification of Ohm's Law. These assertions are supported by their past experience, they claim. Carefully managed discussion can lead DipEds to see that they are reinforcing in a different way the teacher's reasons for this activity — to help them understand and accept the importance of knowing the purpose of a laboratory activity. There is valuable learning in the way in which they show that bringing their own purpose to the

Figure 10.4: A sample response from the V and I experiment

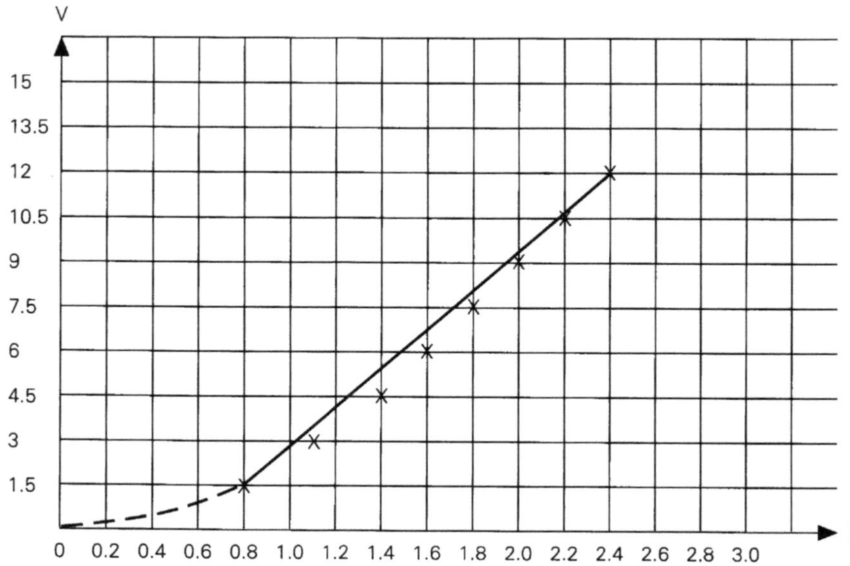

.CONCLUSION: Linear graph

V = IR

we have shown Ohm's Law

. WHAT WAS THE PURPOSE IN DOING THIS EXPERIMENT?

Rather than telling us the relationship between V and current we saw it for ourselves and drew our conclusions that it is a linear relationship
(Plus other comments on translating circuit diagrams to actual equipment)

activity changes the activity. This debriefing discussion takes considerably more time than the actual laboratory work.

This episode differs from what is common in school science in that the DipEds do have an answer to the question about purpose. School students rarely have an answer, except for gross and unhelpful generalities. However the answer here is not always the teacher's purpose, with demonstrable consequences for learning.

An analysis of content in terms of metacognitive outcomes: Again it is familiarity with the content which gives rise to the expectation of Ohm's Law which is central to the metacognitive outcomes, but, once more, it is familiarity mixed with uncertainty. That is, the DipEds do not have a sufficient understanding to be able to take the data at face value, accept the curve, and conclude that some other factor is involved. Also operating here is a perceived familiarity with the reasons for doing laboratory work, a view derived from experience that verification of a law is a likely purpose.

Some Tentative Conclusions

As I indicated at the start of this chapter, this is a tentative first attempt at consideration of links between specific content and metacognitive purposes. Hence I am reluctant to claim too much by way of generalizations from the three examples.

One common feature of the examples is that the content is not completely unfamiliar. The content can be fairly characterized as material which has previously been learned but not understood. This characteristic might suggest that content which is known is appropriate for metacognitive purposes, an argument which does have clear merit in the context of introducing to students new learning strategies involving unfamiliar processes (for example, concept maps). However the issue is more subtle in the case of metacognition. There are powerful arguments (for example, Baird and Mitchell, 1986; Gunstone & Baird, 1988) that the enhancement of metacognition requires content contexts for which real cognitive learning is required of students. Without this, it is very hard for students to see personal value (fruitfulness) in the greater intellectual involvement which the more metacognitive approaches to learning demand. This position, taken in conjunction with the characteristic argued to be common to the examples described in this chapter, suggests that one aspect of content appropriate for the achievement of metacognitive purposes is that it is neither already understood nor totally unfamiliar. The cognitive learning goals within these contexts, which are to be the vehicles for the development of metacognition, must be neither trivial nor too demanding.

The other generalization which can be advanced from the examples is obvious. The nature of the metacognitive goals suggests some specificity about the appropriate content: to develop an appreciation of aspects of observation requires a content context involving genuine observation; to explore the consequences of not understanding a teacher's purpose for a laboratory experiment requires a content context that involves a laboratory experiment. There will also be variation in the extent to which additional specific tasks are needed to assist learners to formulate metacognitive outcomes, as was argued to be necessary in example 1.

References

BAIRD, J.R. (1984). 'Improving learning through enhanced metacognition'. Unpublished Ph.D. thesis, Monash University, Victoria, Australia.

BAIRD, J.R. (1986). Improving learning through enhanced metacognition: A classroom study. *European Journal of Science Education*, *8*, pp. 263–82.

BAIRD, J.R. (1990). Metacognition, purposeful enquiry and conceptual change. In HEGARTY-HAZEL, E. (Ed.). *The Student Laboratory and the Science Curriculum*. London: Routledge, pp. 183–200.

BAIRD, J.R. and MITCHELL, I.J. (Eds.). (1986). *Improving the Quality of Teaching and Learning: An Australian Case Study — The PEEL Project*. Melbourne: Faculty of Education, Monash University, Victoria, Australia.

BAIRD, J.R. and NORTHFIELD, J.R. (Eds.). (1992). *Learning from the PEEL Experience*. Melbourne: Faculty of Education, Monash University, Victoria, Australia.

BARNES, D. (1976). *From Communication to Curriculum*. Middlesex: Penguin.

CHAMPAGNE, A.B., GUNSTONE, R.F. and KLOPFER, L.E. (1985). Effecting changes in cognitive structures among physics students. In WEST, L.H.T. and PINES, A.L.

(Eds.). *Cognitive Structure and Conceptual Change*. Orlando, FL: Academic Press, pp. 163–87.

DE JONG, E.J. and GUNSTONE, R.F. (1988). 'A longitudinal study of some mechanics concepts and conceptual change.' Paper presented at the meeting of the National Association for Research in Science Teaching, Lake Ozark, MO, April.

GAULD, C. (1986). Models, meters and memory. *Research in Science Education, 16*, pp. 49–54.

GUNSTONE, R.F. (1991). Reconstructing theory from practical experience. In WOOLNOUGH, B.E. (Ed.). *Practical Science*. Milton Keynes: Open University Press, pp. 67–77.

GUNSTONE, R.F. and BAIRD, J.R. (1988). An integrative perspective on metacognition. *Australian Journal of Reading, 11*, 238–45.

GUNSTONE, R.F., GRAY, C.M.R. and SEARLE, P. (1992). Some long-term effects of uniformed conceptual change. *Science Education, 76*, pp. 175–97.

GUNSTONE, R.F. and NORTHFIELD, J.R. (1992). 'Conceptual change in teacher education: The centrality of metacognition'. Paper presented at the meeting of the American Educational Research Association, San Franscisco, April.

HART, C. (1987). A teaching sequence for introducing forces to Year 11 physics students. *Australian Science Teachers Journal, 33*(1), pp. 25–8.

MINISTRELL, J. (1982). Explaining the 'at rest' condition of an object. *The Physics Teacher, 20*, pp. 10–14.

POSNER, G.T., STRIKE, K.A., HEWSON, P.W. and GERTZOG, W.A. (1982). Accommodation of a scientific conception: Towards a theory of conceptual change. *Science Education, 66*, pp. 211–27.

RUDD, T.J. (1992). 'Exploring self assessment with primary school students'. Unpublished M.Ed. Studies project, Faculty of Education, Monash University.

WHITE, R.T. and GUNSTONE, R.F. (1992). *Probing Understanding*. London: Falmer Press.

Chapter 11

The Constructivist Paradigm and Some Implications for Science Content and Pedagogy

Malcolm Carr, Miles Barker, Beverley Bell, Fred Biddulph, Alister Jones, Valda Kirkwood, John Pearson and David Symington

Through a comparison of the widely-held traditional view of science with the constructivist view of science, we argue that the constructivist view of the content of science has important implications for classroom teaching and learning. This alternative view of science concepts as human constructs, scrutinized by application of the rules of the game of science, raises many challenges for teachers. Reconceptualization of teachers' views of the nature of science and of learning in science is important for a constructivist pedagogy. We argue here that open discussion of the 'rules of the game' of science would contribute to better learning in the classroom, since learners would be better equipped to change their existing concepts, by knowing more about the nature of science itself.

The Traditional Image of Science

Many teachers hold the view that:

* science *knowledge* is unproblematic
* science provides *right answers*
* *truths* in science are *discovered* by observing and experimenting
* choices between correct and incorrect interpretations of the world are based on *commonsense* responses to *objective data.*

This traditional image of science has been explored in a large number of commentaries; see, for example, Chalmers (1976), Nadeau and Desautels (1984).

Teaching based on this traditional view of science attempts to transmit to learners concepts which are *precise* and *unambiguous*, using language capable of *transferring* ideas from expert to novice (teacher to student) with *precision*. Textbooks are sources of facts and theories about the concrete world, exemplified by the following extract from a textbook about teaching science in the primary school, Jacobson and Bergman (1980):

> Electrical energy can be converted into other forms of energy. As we have seen electrical energy can be transferred considerable distances along conductors. Another great advantage of electrical energy is that it can be readily converted into other forms of energy. (p. 44)

The expectation is that teachers read this material, come to the same understandings as the authors, and then transmit the concepts to their students. The students who learn these statements come to know about electrical energy. There are a number of problems here.

- Study of the ideas students bring to lessons about energy reveals a variety of person-centred views related to 'needing energy to do things', 'needing food as a source of energy', and 'switching off appliances to save energy'. The textbook quote ignores these alternative conceptions, treating energy as a secure and certain idea in a manner which encourages rote learning of phrases with no connection to the learner.
- Energy is a very difficult and complex idea *invented* by scientists. Arguments about the appropriate way to understand and to teach energy continue in education journals (see the debate between Warren, Schmid, Lehrman, Hicks, Richmond and others in *Education in Physics* and *Physics Teacher* in the early 1980s). The textbook quote above treats this area as if there were no difficulty in the concept. Many students can feel that their inability to 'see' the world in the same way as the textbook writer means that they are somehow 'blind' or not seeing the world through the same spectacles.

Teachers who are themselves insecure in their knowledge of science can find the uncomplicated transmission of knowledge attractive, Osborne and Freyberg (1985). Transmissive teaching avoids discussion (since learners lack knowledge worthy of consideration) and interactions which might reveal teachers' uncertain knowledge and so alter power relationships in their classrooms. The view of science as a body of unambiguous right answers for transmission into learners' heads can then trap teachers into a teaching style inimical to their own and their students' learning, or into avoidance of the subject entirely.

A further concern about the consequences of this image of science and good learning is that students who commit the 'facts' to memory are seen to possess a 'natural' ability in science. Others students are given messages that they are not expected to perform as well. Although having a good memory is an undeniably valuable attribute there is a good deal of evidence that the memorized knowledge is not well understood. Teaching which values a skill that may not be strongly linked to ability in science can alienate the bulk of students from the discipline before they have properly experienced it.

Although this picture is extreme, a great deal of current practice in science teaching conforms quite closely to this picture of the interaction between science content and pedagogy. Fensham (1985) has argued that this is unhelpful to learning in science for all students.

An Alternative Paradigm

The traditional view of science and consequently of the language of science has been scrutinized, particularly in the last two decades through considerations of:

- the history and philosophy of science;
- the psychology of learning.

This scrutiny has resulted in science being located in a new paradigm (Kuhn, 1970) in which the nature of science and of learning in science is viewed differently. The new paradigm regards science as a human and social construct, and views learning as the personal construction of new knowledge. As a consequence many of the old securities about the pedagogy of science are seen to be problematic.

Arguments for this new paradigm come from:

- the *history and philosophy of science*. Our perception of the world is seen to be *subjective*. Observations are enmeshed with previous experience of, and existing theories about, the world. We construct meaning for the world around us from our prior attempts to make sense of it. The traditional view of scientific method as the objective scrutiny of hypotheses by reference to unproblematic facts (based on observations), and the consequent rejection of incorrect hypotheses which do not agree with these facts, is now disputed as a description of scientific practice. The alternative constructivist paradigm is less clinical and more human. These are qualities which could well make this view more accessible to young learners;
- *the psychology of learning*. Newer theories see learning as interaction with previously-existing concepts, and as the building of new mental constructs from prior understandings, Osborne and Freyburg (1985), Driver (1988). In these new theories of learning there is increasing recognition of the importance of the *affective* dimension in learning; even in the apparently abstract and objective disciplines of science and mathematics, Claxton (1991). How we feel about the ideas being presented in our learning experiences affects our learning about them. (In this age of concern for animals it would be unwise in the extreme to introduce biology with a programme based on the dissection of small animals!)

Considerations of the importance of *contexts* to learning are allied to this affective dimension. The context in which we learn something affects the way that individuals construct knowledge. Learning about a scientific concept may be much easier through contexts with rich links to students' interests, such as teen culture and the human body, Rodrigues (1992).

A further complication when considering learning in science is the developing realization that individual students hold many, often conflicting, concepts about their world, some of which they use in the school classroom, others in the world outside. Within the chemistry classroom students respond that the world is made up of atoms and molecules, but they talk of materials in their 'real world' in quite different terms, denying that blood and skin and wood are molecular, or even chemicals, Happs (1980). These multiple theories can be used differently depending on *context*. Learners of science who have categorized humans as animals will enter a butcher's shop displaying a sign 'No Animals Allowed' without demur (Bell, 1981). A significant challenge to teachers is to make the science learned in school appropriately available in the real world of the student. Much of the discussion about enhancing learning of science seeks ways to make the content of science more plausible, intelligible and fruitful for learners, Hewson and Thorley (1989), thus helping them to make better sense of their world. Whilst accepting that this is a very important activity this chapter argues that some examination of the process whereby science concepts are constructed is also important. The way that constructs are scrutinized, in other words the rules of the game of science, deserves consideration in teaching and learning. If science is a collective construct then exploration of the constraints on this process of construction deserves to be part of classroom interactions.

Constructivist Teaching Approaches

If students come to lessons with ideas about their world which already make sense to them, then teaching needs to interact with these ideas, first by encouraging their declaration and then by promoting consideration of whether other ideas make better sense. These procedures have been outlined in the generative teaching approach of Osborne and Freyberg (1985) and the interactive teaching approach of Biddulph and Osborne (1984) and further discussed in Driver (1990).

A feature of this approach to science teaching is that the outcomes can be different for different students. Some may want to explore a concept in considerable detail and will develop understandings closer to those of the scientist, while others will be more interested in exploring practical and personal aspects of the topic. This diversity of outcome poses problems for teachers. The outcome from traditional science lessons is also diverse, though assessment procedures that rely heavily on recall and rote learning conceal them. When understanding is probed at a deeper level the learning is often found to be superficial, even for students who have been described as very successful. The problem for learners who *are* described as successful is that they are often unaware of the partial nature of their development of a particular concept, and find difficulty in contemplating change to their ideas. Procedures in which there is more conversation about learning provide a better base for further learning. The open negotiation of meaning, and appreciation of the partial nature of the learning achieved, also model a better image of science.

The Process of Developing Science Content in Constructivist Teaching

Rather than discussing constructivist teaching in a generalized manner the remainder of this chapter refers to three topics, *floating and sinking, energy* and *photosynthesis*, and considers some possible references to the process of science which could be introduced when these are the focus of learning in the classroom. The discussion draws on research into teaching and learning about these concepts, particularly that reported in Biddulph (1983), Osborne and Freyberg (1985) and

Carr, Hayes and Symington (1991) on floating and sinking; in Carr and Kirkwood (1988) on energy; and in Barker and Carr (1989) on photosynthesis.

Floating and sinking is an appropriate topic for detailed consideration since it is a common topic in primary science programmes, often assumed to provide no particular problems for learners (the New Zealand Minister of Education recently stated that knowing that heavy things sink is a significant beginning to early learning in science). Energy is another concept introduced to primary school students; the argument being that a fundamental concept in science should be introduced in a simple manner at an early stage in learning. Photosynthesis is also seen to be a fundamental concept, most commonly introduced through the notion of plants using energy from the sun to make their own food; a notion that has a number of difficulties for learners.

Teachers reflecting on their pedagogy could usefully consider five questions as they develop their teaching approaches and reflect on student learning. These are:

- Does nature contain a definition of the concept which can be uncovered through appropriate experiences?
- How does science develop a statement of a concept?
- Is there a single explanation for a phenomenon which teachers should aim at?
- Can science always provide an answer to a question?
- When a 'better' explanation is proposed how do scientists decide to accept it?

These five questions raise a number of important issues about the concepts in science and consequently about science pedagogy. The following sections consider each in turn.

Question 1: Does nature contain a definition of floating and sinking, energy, and photosynthesis which can be uncovered through appropriate experiences?

The implied view of much science writing in texts and teaching is that a word such as floating has a *meaning which exists in the world*. This meaning is seen to be independent of people, unambiguous, and apparent to the trained and careful observer. Through a process of exploration of their world, learners are expected to unearth this *true meaning* of the word, provided that a teacher guides the process with skill.

The word floating does not, however, seem to have exact meaning derived unambiguously from experience. This is apparent when we look at some instances (figure 11.1). The first example, the apple, may seem obvious and uncomplicated. Surely the apple *is* floating. Yet some young people will say that the apple is partly floating and partly sinking (which makes sense if your idea of floating is to be above the surface of the water and of sinking is to be below that surface). If the illustration of the stones dropped into water showed one of the stones partly above the surface then the logic of this view could be that some of the stones are sinking but that one is floating. Consider the person under the water looking at fish. Is the person floating? If our concept of floating is that something is to be *above the surface of the water*, then the answer is no. If we know that an apple pushed under water will bob up again when we release it we may decide that

Figure 11.1: Instances and non-instances of 'floating'

floaters can be submerged temporarily, but that they remain floaters. How would this affect our response to the survey question about the underwater swimmer? We know that divers can swim back to the surface, unless something goes wrong (and the person illustrated does not look distressed). By this argument perhaps the person is floating. Another view of floating involves the common experience of feeling *supported by the water*. From this viewpoint the diver is floating, even although submerged. The same considerations apply to the submarine. Further complexities arise when we turn to the yacht in trouble. In our normal use of language the yacht is sinking but in the illustration it is still floating, if only just! The person swimming may set off different associations depending on our own experiences. For some people water is a dangerous environment, raising fears of drowning. This illustration then shows a person who is going to drown if he/she stops swimming. Other responses may arise from a personal experience of comfort in the water, knowing that when you cease swimming you can float. Each association could lead to a different response. The person swimming has introduced a puzzling new dimension, that of movement. The same problem applies to the stone skipping along on top of the water. We know that most stones would sink if they stop moving (in a country like New Zealand where pumice is often found there are stones which float!). Does that mean that they are floating? The same difficulties apply to the speeding boat.

Complications seem to be piling up endlessly but the point to make can now be stated clearly. The idea of floating does not exist sharply defined in nature. We need to construct a meaning for it — which is just what scientists have done. Their concept of floating and sinking refers to many experiences, and tries to make their observations more coherent. In the process they have invented concepts which at times they have agreed to change. This is a good description of science, and one we argue should be shared with learners as their learning in science is developing.

Similar considerations apply to the concepts of energy and photosynthesis. As learners (both students and teachers) explore their concepts, they will find that there are a number of conflicting ideas (concepts of energy may centre on food, and on people having or requiring energy). Conservation of energy is a difficult construct when prior ideas strongly support the need for continual supplies of food to keep one's energy up, and, paradoxically, when people exercise vigorously to build up their energy! Prior ideas related to photosynthesis are also likely to require consideration. The concept of plants making their own food from the energy of the sun contains many pitfalls, particularly when it is commonly held that plants get their food from the soil, and the idea of invisible gases providing the substance from which plant material is constructed has so few links to prior experiences.

There are, then, no definitions of these concepts to be uncovered by appropriate experiences. Rather each construct is gradually made from many instances, and its power is dependent on bringing together many related ideas.

Question 2: How does science develop a statement of a concept such as floating and sinking, energy, or photosynthesis?

The survey on floating and sinking provides clues to the answer to this question. Development of a construct of floating first confronts the complexity of our rich world of experience. Scientists, for very good reasons, aim to reduce this richly complicated world of experience to a more manageable, tidied-up one. For science to 'work' concepts are sharply focused on aspects chosen to be relevant and reproducible. Paradoxically this process leads to concepts which have very great power in making predictions about the world, and this is the 'pay-off' which has made science one of the great human inventions.

In the case of floating and sinking the scientists' construct first seeks to make these two possibilities mutually exclusive (something is *either* floating *or* sinking, and ambiguities such as partly floating and partly sinking, floating now but about to sink, are avoided). The iceberg, then, is viewed by science as floating, even though many people will tell you that the part below the water is sinking and the rest is floating. The *total system* is defined as an instance of floating. By reducing the complexity of the real situation, simpler and tidier statements can be made. The person in a life jacket provides a further illustration. We might think that the person is sinking and the life jacket provides support so that the sinking person can float. Again by treating the system as a whole, science chooses a simpler, more reduced view of the world. So the first strategy is to treat systems as a whole. A similar decision about defining a system needs to be taken when energy is the focus of teaching and learning. Many of the problems associated with this topic are clarified by clearly defining a system and a change to that system. Students were often unsure whether to remove the human operator from systems. Switching on a light implied the use of muscles, and before objects could be dropped they had to be raised or pushed by a person, Carr and Kirkwood (1988).

The next decision about floating is to shift the focus from the surface of the water. The new construct considers something to be floating if it is *supported* by the water (even if it is under the water). This extends the concept to more situations. The yacht in trouble is therefore considered to be floating since it is still supported by the water. This example provides a further instance of the way that meanings for concepts are constructed in science. The situation is treated *as it is at the moment*, regardless of what may happen next or what may have happened before. Science often takes this view of the world, disregarding the past history of an event, and any possible future. This provides a sharper focus, even though information from our prior knowledge is now ignored. Of course, in everyday language the yacht *seems to be* sinking (though some can float like this for days), but the scientists' construct is that at the moment it is floating. It is valuable to think about how this construction of meaning does not always remove some difficulties. Consider now the empty plastic bottle tied to a rock. The bottle is described by science as *floating* although many people would describe it as sinking or sunk since it is beneath the water. Is the scientists' response based on the understanding that if the bottle is released it will float to the surface? This is not the case. The scientist is not using knowledge of what might happen in the future. The decision is based on seeing the bottle as being supported by the water (experiencing an upthrust) and therefore *floating*. This reduction to a simpler, more extended concept is typical of the process of science. Indeed the concept of floating and sinking has been extended by science to liquids and gases other than water. Focussing on the moment and using the concept of support provides a more powerful idea. By this process science has been developed as an interrelated set of constructs about the world, in which carefully defined ideas work together to inform us about much of our experience of the physical world. When scientists decide to confer onto a word a particular, more exact definition this is an act of construction from the world of experience significant to scientists. The process of teaching and learning ought to engage with this 'rule of the game' and its consequences.

The concepts of energy and photosynthesis have also been constructed by selection and reduction from the world of experience and of ideas, and their modern meanings are powerful because of this. Realizations that different phenomena such as heat, electricity and mechanical work could be quantitatively interconverted led to the invention of the concept of energy and of its conservation. Photosynthesis is an elaborated concept which depended for its clarity on unravelling of the concept of gases, as well as the development of the concept of energy. The power of the ideas has a great deal to do with their connectedness, which is why teaching can not expect successfully to develop the concepts from a single experience or a few isolated experiences. Neither can teachers expect to help students change their concepts without deliberate explorations of their connectedness.

In the process described above, science has often reinterpreted experience and as a result the words of science have changed their meaning. Consider the word 'atom'. This word has been used in science since the ancient Greeks. Their concept of atom held that all substances were formed of their own particular atoms (there was an atom of milk, an atom of blood, an atom of glass and so on). Each substance was made up of a large number of similar atoms; the concept implied that there was a very large number of different atoms. Later the word atom came to mean an unbreakable unit of matter, and substances could be compounded of several different atoms. The number of possible different atoms was consequently greatly reduced. It was difficult to decide how many distinct atoms there were, so elaborated rules were constructed to decide this. Eventually the construct, atom, became very sharply defined and scientists could state that all materials were made from a defined number of them. Since that time of certainty the picture has clouded. The discovery of isotopes and of elements beyond uranium required that the construct be changed. More recently quantum theory has necessitated a further reconstruction of the word. Science no longer defines atoms as the indestructible building blocks of matter since many sub-atomic particles have now been defined. The point being made here is that the concept atom does not exist out in the world somewhere waiting to be discovered; the word is undergoing a process of continual revision of meaning in the light of further explorations of the world.

This reduction of the world to manageable aspects should be understood for what it is. Rather than science being thought of as a 'given' feature of the world around us, the learner needs to appreciate that science has been *constructed by people* as a way of making better sense of the world.

This has obvious implications for science teaching and learning. If science develops by taking decisions about what would make better sense of the world, then teaching approaches could describe this procedure candidly and at a level appropriate to the learners' knowledge at the time.

Scientists, then, deal with an idea such as floating and sinking by tidying up the world and then inventing powerful ideas to make better sense of it. This process involves conscious building of new ideas from prior knowledge. There is a tension here between reducing the world to 'manageable' simplicity and then treating systems as a whole. A vital debate about the difference between Western science and other means of knowing would be informed by students being aware of this tension, Christie (1991).

Question 3: Is there a single explanation for the phenomena of floating and sinking, energy, and photosynthesis which teachers should aim at?
A common view of science is that there is a single scientifically acceptable description or explanation which teachers need to instil in their students. This is frequently apparent when teachers talk about how they should never tell their students anything that is wrong, and that the end result of any sequence of lessons is to produce the right answer. This encomium places an impossible burden on teachers, one which some are not aware of. School science can only be provisional knowledge leading towards the scientist's construct. In most cases this scientist's concept is inaccessible to students but transitional concepts can be valuably addressed. Preparedness to acknowledge the provisional nature of their learning and to accept that future change will be necessary may be a very valuable part of science teaching and learning. If we consider the concept of floating and sinking, the following explanations of why an apple floats are useful provisional concepts for learners from which the scientist's concept could be developed.

The apple floats because:

- it is light for its size
- it is held up by the water
- the upthrust of the water is equal to the weight of the apple
- the weight of the water displaced is equal to the weight of the apple
- the apple's density is less than the density of water
- the difference in water pressure below and above the apple results in a net upward force equal to the weight of the apple.

The level of explanation depends on the purpose of the exploration and the background of the person for whom explanation is being provided. It is inappropriate for classroom interactions to convey the impression that there is a single correct explanation of any phenomenon or a single definition of any concept. This is a challenge to the pedagogy of most teachers.

Question 4: Can science always provide an answer to a question?
There are at least two instances in the survey which test the scientific view of floating and sinking to the limit; the spider standing on the surface of the water, and the stone skipping on the water. Indeed, these exemplars remind us that in the current description of phenomena 'floating' and 'sinking' are inappropriate terms for their analysis. The two instances show a water surface and objects associated with it, yet we have real difficulty in answering the question 'Is the object floating?' and *so does the scientist.* Some scientists would say that the spider is floating because it is supported by the water. Others would say that the phenomenon involved here is surface tension and that the spider is not floating but rather supported by a 'skin' on the top of the water. We can support a needle on a water surface by carefully lowering it onto the surface, but this needle will sink if the surface is disturbed or some detergent is added to the water. The concept of floating held by most scientists requires that water is displaced, but in the case of the spider this may be said not to have happened.

What about the stone skipping on the water? This is a tricky one since the movement of the stone is vital for it to remain on top of the surface. When its speed drops sufficiently the stone will sink. Science *cannot answer the question* 'Is the object floating?' in the spirit in which it was asked, even though it looks like a perfectly reasonable one. (There are similar problems with the speedboat, though in this case the boat may well float if it stops moving). The scientific idea of floating and sinking is embarrassed when objects are in motion because the idea then becomes too complex for a straightforward analysis.

The implications of this for teaching are clear. Young learners are very likely to ask questions which do not fit tidily into the constructs of science even though they seem simple. The apparent simplicity of the language conceals a very difficult problem. A teacher who understands that science does not have all the answers may feel more comfortable about helping a student to explore the world than the teacher who feels threatened by challenging questions. It should not be considered unprofessional for a teacher to acknowledge that some questions are unanswerable at the level of knowledge of science of the learner. This issue becomes clearer through some other examples of apparently simple questions, like: 'Is there more energy in a glass of milk than a glass of water?' The proper answer to this question is that there is insufficient information for an answer to be given. We need to know what happens next. As already discussed the nature of science is to exclude from the system past history and future possibilities. In this case there are a number of possible futures. If the glasses are drunk then one analysis applies; however if they are thrown at something then the analysis changes. The question can be answered but only when more information is provided. Again a simple question, posed in apparently exact terms is one which cannot be answered. Questions like this can be good questions for exploration of basic understanding.

The last question comes to the heart of the construction of ideas in science, since it explores the process of change in the constructed meanings of ideas.

Question 5: When a 'better' explanation is suggested, how do scientists decide whether to accept it?

If science is not a set of truths which exists independently of people, then in the construction of this structured complex of ideas there will often need to be *changes* made to ideas. This process of changing prior ideas is also the core activity of education so the issue of acceptance or rejection of a new idea is an important one both for science and for science education. We now know that learners often retain their prior meanings for worlds rather than taking on new meanings and we also know that this situation is often not apparent to the teacher. If your idea of floating was based on an object being above the water surface then introduction of the 'better' idea that floating involves support by the water will have caused problems. You would need to have explored the new idea and found it to be more helpful and to offer better explanations before you would feel comfortable with it.

Scientists have frequently had to face this difficulty. Although a popular image of scientists is that they quickly accept new discoveries the history of science shows that many scientists have continued to disagree with developments in their field for very long times (one commentator on the history of ideas has remarked that new theories in science are finally accepted when the last opponent dies of old age). When an idea has become part of the way that you think about the world, changing that idea will mean overcoming barriers which can be very difficult to break down. Nevertheless there are some 'rules of the game' of changing ideas that can be offered. They have to do with new ideas being:

- more parsimonious (if a single explanation encompasses several others then science tends to prefer the more inclusive);
- more elegant and coherent (if the new idea is tidy and links to a greater number of other ideas then it will be preferred);
- able to explain what is already known and also to predict some unknown outcomes. Ideas are preferred which have explanatory and predictive power.

There can be valuable debate over these statements. The point made here is that addressing them would assist learners and teachers to explore the process of the construction of scientific knowledge. Much of the current critique of science and its products does not confront the nature of science and the procedures used in its construction. The best criticism comes from understanding these rules, followed by an analysis of their consequences. Many of our students reject science without any understanding of the power of its constructs and the manner by which these have been constructed.

The important point for our discussion here is that when teachers are encouraging students to adopt a *different* construction of a concept, classroom activities require more than statements which outline the new meaning, perhaps followed by some exercises. Learners need time to consider their prior meanings, to explore new ideas, to link them to other existing ideas, and to construct new meanings knowing what the requirements are for this reconstruction. The false view of science which holds that scientists are rapidly converted to new ideas has resulted in classroom approaches to science which too often assume that changing ideas is an easy procedure. The most important feature of an approach to science classes which addresses the difficulty of changing ideas is *conversation*. Science lessons which continually seek learners' ideas, which help to clarify them, and which provide an open and unthreatening environment for changing these ideas through conversation are classes in which learning in science can be improved. The false idea that science is exact and therefore that concepts in science are unproblematic can be argued to have trapped science teaching into a pedagogy which misrepresents both the content of science and the process whereby this content is constructed.

Summary

This chapter argues that classroom teaching and learning should address the processes of constructing and reconstructing scientific knowledge. A number of issues arise naturally in the classroom which enable teachers to explore the nature of science. These include:

- how by building up experiences and trying to make better sense of them scientists construct concepts — which are not 'out there' in the world waiting to be uncovered;

- how science takes the richly complex world of experience and reduces it to a more manageable one in order to make more powerful statements;
- how science often focuses on a defined system as it is at the moment and ignores previous history or possible futures unless these are clearly stated;
- that there are a number of apparently simple questions about the world that may be very difficult for science to answer;
- that the rules for deciding that a proposed explanation is better include notions of elegance, parsimony and greater connectedness as well as those of plausiblity, intelligibility and fruitfulness. The process of making a substantial change to the ideas of science is one which has aroused and continues to arouse argument and even passion.

References

Barker, M. and Carr, M. (1989). Teaching and learning about photosynthesis. Part I: an assessment in terms of students' prior knowledge. *International Journal of Science Education*, *11*(1), pp. 49–56.

Bell, B. (1981). When is an animal not an animal? *Journal of Biological Education*, *15*(3), pp. 213–18.

Biddulph, F. (1983). *Students' Views of Floating and Sinking*. Working Paper No. 116. Hamilton, New Zealand: Science Education Research Unit, University of Waikato.

Biddulph, F. and Osborne, R. (1984). *Making Sense of Our World: An Interactive Teaching Approach*. Hamilton, New Zealand: Science Education Research Unit, University of Waikato.

Carr, M., Hayes, D. and Symington, D. (1991). Language and science: constructing a sense of the world. In Furniss, E. and Green, P. (Eds.). *The Literacy Connection; Language and Learning Across the Curriculum*. Melbourne: Eleanor Curtain Publishing, pp. 79–98.

Carr, M. and Kirkwood, V. (1988). Teaching and learning about energy in New Zealand secondary school classrooms. *Physics Education*, *23*, pp. 86–91.

Chalmers, A.F. (1976). *What is this Thing Called Science*? St. Lucia, Queensland: University of Queensland Press.

Christie, M. (1991). Aboriginal science for the ecologically sustainable future. *Australian Science Teachers Journal*, *37*, pp. 26–31.

Claxton, G. (1991). *Educating the Enquiring Mind: The Challenge for School Science*, Hemel Hempstead: Harvester Wheatsheaf.

Driver, R. (1988). Theory into practice II. In Fensham, P. (Ed.). *Development and Dilemmas in Science Education*. London: Falmer Press, pp. 133–49.

Driver, R. (1990). *Constructivist Approaches to Science Teaching*. Seminar paper presented at Mathematics Education Department, University of Georgia.

Fensham, P.J. (1985). Science for all: A reflective essay. *Journal of Curriculum Studies*, *17*, pp. 415–35.

Happs, J. (1980). *Particles*. Working Paper 18. Hamilton, New Zealand: Science Education Unit, University of Waikato.

Hewson, P. and Thorley, R. (1989). The conditions of conceptual change in the classroom. *International Journal of Science Education*, *11*(5), pp. 541–53.

Jacobson, W.J. and Bergman, A.B. (1980). *Science Activities for Children*. Englewood Cliffs, NJ: Prentice Hall.

Kuhn, T. (1970). *The Structure of Scientific Revolutions*. Chicago, IL: University of Chicago Press.

Nadeau, R. and Desautels, J. (1984). *Epistemology and the Teaching of Science.* Ottawa: Science Council of Canada.

Osborne, R.J. and Freyberg, P.F. (1985). *Learning in Science; The Implications of Children's Science.* Auckland: Heinemann.

Rodrigues, S. (1992). D.Phil. study in progress, personal communication.

Chapter 12

Making High-tech Micrographs Meaningful to the Biology Student

James H. Wandersee

The Problem

Context

As we enter the twenty-first century, biology texts and classroom laboratory exercises are making increasingly greater use of high-tech microscopic images (micrographs) in teaching important biological concepts in topical areas ranging from cytology to systematics. Indeed, up to a third of the total pages in today's introductory biology texts may be allocated to graphics (Frazier, 1991). Yet, research shows that introductory biology students frequently bypass the micrographs in their textbooks because they find them incomprehensible and because their understanding of micrographs is seldom tested on biology examinations (Frazier, 1991; Nist and Kirby, 1989). If, as Novak and Gowin (1984) argue, 'to learn meaningfully, individuals must chose to relate new knowledge to relevant concepts and propositions they already know' (p. 7), then we must *promote* such choosing and attempt to bridge the gap between students' current knowledge and the new knowledge we would like them to learn.

Recent research by Chan, Burtis, Scardamalia and Bereiter (1992) into learning from texts supports this sense of the learner actively 'choosing to relate . . .'. The learners' prior knowledge was an important variable, but the activity of the learners in the process had a direct and significant effect on learning.

Complicating Factors

The living microworld, with all its intricate microstructures and ultrastructure, lies beyond students' everyday experiences (Burgess, Marten and Taylor, 1987; Wandersee, 1981 and 1986). Extrapolating knowledge of the macroworld to the microworld is seldom helpful. Students often do not see or recognize discrete structures in a micrograph and they report that puzzling over the micrographic maze of visual complexity gives them a headache. In addition, we know from alternative conceptions research that students often dismiss what they cannot directly observe as non-existent — as if it were just something that scientists 'make up' (Mintzes, Trowbridge, Arnaudin and Wandersee, 1991).

The problem with the typical captioned micrographs found in today's biology textbooks is that they allow only those who already understand that microworld to 'see' it as biologists do. To novices, the same micrographs are but a small, 'knowledge-fogged' window on that microworld. In some ways, publishers tacitly acknowledge this perplexing situation and, in doing so, perpetuate the conceptual marginality of micrographs by placing them outside the textstream and by typesetting their captions in smaller type. Thus, micrographs often serve as a visual footnote to the text, a minor and tangential comment to be studied only by the few who enjoy biological arcana and have the time for such pursuits.

The Missing Content Pedagogy

There appears to be a significant mismatch between current pedagogical practice and the microstructure-based content of biology, a mismatch that impedes or blocks meaningful learning in such topic areas. Students have enough trouble understanding the images they see with their laboratory bench microscopes, let alone those that originate in the exotic imaging systems of biological research (for example, the scanning electron microscope [SEM], the transmission electron microscope [TEM], and the computer-enhanced light microscope [CELM]).

Shulman (1987) claims that content knowledge and pedagogical knowledge must be interwoven in teaching to create a new type of knowledge, 'pedagogical content knowledge'. In this chapter I argue that biology education needs a pedagogy of high-tech microscopy, and I attempt to begin construction of some salient propositions and teaching strategies which such a pedagogy requires. Although I have conducted research projects in microscopy education, and have also taught biology at the high school and college levels across a span of twenty-one years, I am keenly aware of Hoz, Tomer, and Tamir's (1990) warning that pedagogical content knowledge (PCK) in biology is difficult both to set forth and to explain — because it is often tacit knowledge that is self-constructed by teachers. In addition, I have not been able to find an organized body of research findings in this domain.

The Challenges of Teaching Biology with High-tech Micrographs

Biology educators must recognize that knowledge which is to be constructed chiefly on the basis of high-tech micrographs requires a specialized pedagogy. Not only are these two-dimensional images non-trivial spatial, chemical, and visual transformations of the actual three-dimensional objects which they represent, Weakley (1981) contends that an electron micrograph is 'a photograph of the shadow of an artefact' (p. 124). Most students appear to require perceptual cuing if they are to 'see' what their biology teacher wants them to observe, much less to understand what those regularities mean in the context of biology. When gazing at the world of life through the window of high-tech microscopy, students frequently lack the critical prior knowledge and experiences to which the biology teacher might anchor the lessons he/she wishes to teach (Henk, Territo and

Wandersee, 1992). In addition, students often fail to recognize (or even 'see') the patterns within such micrographs to which their teacher often refers when he/she is attempting to help them make biological meaning of what they are viewing.

Linking Perception to Concept Formation

Recent relevant research was conducted as part of a project entitled *Exploring Microstructures: Introducing Students to the Images, Tools, and Applications of High-Tech Microscopy.* After collecting a large set of data on how introductory college biology students interpret text-based TEM micrographs (the type that students find most difficult to analyze) of cells, it became obvious that students seldom perceive the patterns within the micrographs (which they must recognize in order to form microstructure-based concepts) unless the lesson is carefully crafted to guide their observation (Wandersee, 1992a). Just as a physics student will not 'see' the rolling ball on his/her inclined plane as a sphere moving on a tangent unless taught to do so (Matthews, 1992), so the biology student will not 'see' cells and their nuclei without guidance from the teacher.

The prokaryote-eukaryote distinction is often poorly explained in biology textbooks but is fundamental to understanding a number of key biological constructs and principles — from classification to ecology to cellular evolution — since the distinction also makes use of TEM micrographs, arguably the most intellectually challenging for beginners to interpret (Burgess, Marten and Taylor, 1987, p. 10). I used this teaching problem as a test case for inferring preliminary principles of a content pedagogy centered upon the images of high-tech microscopy. (More information on this research project will be supplied in a later section of this chapter).

Applying Learning Theory

My research is based upon the Ausubel-Novak-Gowin theory of learning (Ausubel, Novak and Hanesian, 1978; Gowin, 1981). As Novak and Gowin (1984) point out, concepts are patterns or regularities in objects or events. Once the student sees the visual pattern which his/her teacher wishes to emphasize in a set of micrographs, the current biological term (concept label) can be attached to that pattern, and the meaning of that concept can be constructed by connecting it to the previously learned meanings of related biological concepts. Gradually, a network of concepts is erected in what one researcher calls 'psychological space' (Fisher, 1990) and the student's understanding gradually expands as the number of relations between the new concept and other concepts increases. Thus, the aforementioned educational problem is considered to be amenable to conceptual analysis using learning theory.

From Low-tech to High-tech Microscopy

If, as Shulman (1992) has proposed, the assumption that 'the history of the concepts in a field is a chronicle of the pedagogy in that field' is indeed valid, then it will be instructive to return to the origins of microscopy to see what we can learn from the histories of science. Because I am, by avocation, a Leeuwenhoek scholar, I begin by examining the work of the 'father of microbiology' in order to search for pedagogical insights.

Leeuwenhoek's Microscopic Explorations

Background information

The Dutch amateur microscopist Antony van Leeuwenhoek (1632–1723) has been called one of the world's greatest biologists, the greatest of the classical microscopists, and one of the most extraordinary self-taught scholars of all time (Crockett, 1978; Gray, 1961; Singer, 1950). During his lifetime, he ground 419 microscope lenses and made 247 single-lens microscopes (Dobell, 1932). While investigators of his day used compound microscopes, he saw more of the living microworld with his carefully crafted single-lens instruments than other microscopists would see a century later.

Although a Leeuwenhoekian microscope lens had a restricted field of view, had to be held close to the eye of the observer, required precise illumination of the specimen, and could not be used without a mechanical stage, it eliminated much of the spherical and chromatic aberration plaguing the compound microscopes used by his contemporaries (Wandersee, 1980). Thus, he made the first research-quality microscopes (designed to optimize both resolution and magnification power) and he went on to become the great investigator whom some historians would call the father of bacteriology, protozoology, parasitology, hematology, and histology. The microscopic images he saw were converted to detailed line drawings (by his personal draftsman) and became part of his 112 scientific reports published in the *Philosophical Transactions* of the Royal Society of London and his twenty-six reports published by the Paris Academy of Sciences (Gray, 1961).

Leeuwenhoek's images

Some of the most important organisms Leeuwenhoek discovered were bacteria. In his book, *The 100, A Ranking of the Most Influential Persons in History*, Michael Hart (1978) ranks Leeuwenhoek thirty-ninth, just ahead of Plato, who ranks fortieth. Hart justifies this rating by noting that Leeuwenhoek discovered microbes and the scientific world built on his discoveries in fighting the diseases that affect millions of people. Leeuwenhoek's discovery of bacteria, Hart contends, was at the time totally unanticipated and independent of other scientific research. How do we know that he really saw bacteria almost a century before any other microscopist did? Hardin and Bajema (1978) list three types of visual evidence: (i) he produced accurate and carefully shaded drawings which can be analyzed; (ii) he measured bacteria by comparing their size to the eye of a louse, a human red blood cell, and a sand grain (and he even made drawings to illustrate his micrometry); and (iii) he described the motion of these organisms (which allows today's bacteriologists to narrow the possibilities for organisms which may be depicted in the sketches). Even without using histochemical technique (for example, staining), Leeuwenhoek provided enough accompanying information so that his images could be interpreted across the centuries.

The relationship between instrument, image, and understanding

Just as Galileo's telescope and its images opened humankind's eyes to outer space, Leeuwenhoek's microscopic images revealed another new frontier — inner space (Wandersee, 1981). If it is true that one can never really interpret the results that a scientific instrument yields unless one understands both how the specimens are prepared for it and how the results are generated by it, then Leeuwenhoek's success is partially explained. By making his own instruments from raw materials, by preparing his own specimens for viewing, by developing his own mechanical stage and method of illumination, he was in an ideal position to understand both the strengths and the limitations of his instrument, as well as the images it produced.

Important Changes in Microscopic Representation as Microscopy Developed

When physicist Robert Hooke (1665–1703) published his famous book *Micrographia* in 1665, it created the first widespread interest in the living microworld, as was Hooke's intent (Ford, 1985). In it, he presented many of his own detailed drawings (in the form of engravings) and Ford thinks that even Leeuwenhoek was inspired by these drawings and writings.

As the art of biological illustration improved, more and more scientific information about the living microworld could be encoded in the art work. Even with the availability of photomicrography, continuous tone drawings are still used today for educational purposes because the illustrator can do things that the camera can not do, such as change the viewing perspective, show subsurface relationships, decide to eliminate extraneous detail, or portray a series of events in a single drawing (CBE Scientific Illustration Committee, 1988). Rather than simply documenting microstructure as a camera might, illustrators can alter a microscopial rendering to enhance understanding.

The Evolution of Photomicrography and the Biological Image

Darius (1984) points out that simply coupling an early camera to a microscope did not necessarily yield a valuable scientific photograph, even though 'medicine and astronomy were (ultimately) the first two scientific disciplines to benefit from the application of photography' (p. 22). In fact, some historians of science contend that photomicrographs were initially less important for scientific research itself than for communicating the findings of science in journals. For example, not until the 1880s did the noted bacteriologist Robert Koch admit that micrographs were very useful in settling scientific debates among researchers and that they were sometimes more important for scientific work than possession of the actual specimen.

Advances in sectioning and staining compensated for some of the previous advantages of illustrations over photomicrographs, and new color-balanced films overcame the camera's unrealistic, differential response to various colors of light. Subsequent progress in micrography depended more on the development of new imaging instruments than it did on new cameras or film. A wide array of microscopes forming images using various forms of electromagnetic radiation (for example, ultraviolet light, electrons) was invented to meet a variety of research demands. With respect to the imaging process, as instrumentation improved, the image produced moved a greater number of data transformations away from the original specimen. In addition, advances in electronic computing power allowed computer reconstruction and enhancement of any micrographic image. (As manipulability of the image grew, micrographs were less likely to be considered incontrovertable evidence). From the basic imaging system consisting of Leeuwenhoek's single-lens microscope held next to the observer's eye, we have progressed to multi-million-dollar, high-tech imaging systems that occupy a small room and can generate/encode images on cathode ray tubes, magnetic media, or photographic emulsions. Living specimens are often incompatible with such imaging systems; not only must the specimen be dead, but it must also be coated with a thin layer of metal. Paradoxically, you could say our images have moved both closer to (in detail) and farther from (in data space) the reality of the living microworld. Thus, Weakley's (1981) observation that a TEM image is 'the photograph of the shadow of an artefact' — greatly enlarged — captures this 'cost-benefit' relationship.

Derived Pedagogical-Content Hypotheses

I return now to the issue of pedagogical content. From my own classroom teaching experience, observation of other classrooms, biology student and teacher conversations, and data resulting from collaborative, classroom-based investigation of common barriers to understanding high-tech images, I infer the following pedagogical 'principles' for the reader's consideration in improving this aspect of biology teaching. These are not 'commandments', but simply assertions that I think are worthy of further consideration (i.e., hypotheses).

1 Teachers should plan a variety of simulated and direct *experiences* with the living microworld (for example, cell ultrastructure modeling, computer simulations of cells, cell-based board games, pond water video-microscopy).
2 Biology textbooks should carefully *integrate* high-tech biological images with the text itself — giving them equal status with words.
3 Students should actually be *tested* on their interpretation of high-tech micrographs.
4 Students should move from direct observation with a single-lens magnifier, to observation (and then Polaroid™ photomicrography) with the light microscope, to interpretation of simpler high-tech biological images (for example, CELM, SEM) to interpretation of more complex high-tech biological images (for example, TEM), *moving deliberately from direct to vicarious experiences.*
5 A content pedagogy for micrographs could employ Ausubelian *advance organizers* to explicitly help students move from what they already know about the living microworld to what they are going to learn.
6 Students should be guided by the teacher to perceive the *visual patterns* in micrographs that underlie key concepts in biological microstructure

before the students are told the term (concept label) or taught the function of that microstructure. (Seeing is not *ab initio*).

7 Seeing the visual patterns in biological micrographs seems to require that the student *examine a representative set of micrographs* depicting the concept of interest, not just a single example of that concept.

8 The *history of microscopy* can and should serve as a pedagogical heuristic. (We did not leap to the present technology, but used earlier, simpler technologies first.)

9 The *strengths and weaknesses* of the major types of microscopy-based illustrations and micrographs should be taught. The choice of an imaging system should be seen as context dependent and theory laden.

10 Leeuwenhoek's imaging strategy of (i) pictorial documentation, (ii) micrometric analysis, and (iii) time-series representation should be used to generate *guidelines for micrograph usage.*

11 Students should be introduced to *simpler* (or selective, feature-enhanced) versions of the actual unaltered biological micrographs that they will later have to interpret.

12 Students should learn *how specimens are prepared* for each biological imaging system and *how each imaging system works* (basic theory of the instrument).

13 Students should be taught to identify the instrument which *made* a particular high-tech biological micrograph by carefully examining a set of its micrographs and learning the image characteristics for each of the common high-tech microscopes.

14 Students should understand when it is most appropriate to *use* a particular biological imaging system and what the *trade-offs* are with respect to realism, resolution, and magnification.

15 Students should be taught how to *mentally reconstruct* the original specimen from a series of sectional views and to *apply the basic principles of micrometry* in analyzing a micrograph. They should then be shown computer image analysis programs which can now perform similar functions.

16 Students should study and learn the basic *functions* of micrographs as sources of evidence in biology.

17 Students should discuss how high-tech biological images (for example, TEM, SEM) differ in their *correspondence to reality* from low-tech ones (for example, LM).

18 Students should know how the *wavelength of the radiation* which produces the image affects a microscope's *resolution limit.*

19 Students should realize that, because it is now so easily modified, today's high-tech micrograph does *not* always serve as incontrovertable evidence in science.

20 Students should explore the principle that computer enhancement of an image involves the electronic processing of an image file to *improve the signal-to-noise ratio, alter contrast, remove background clutter,* or *add pseudocolor* (since all original electron micrographs are in monochrome). Modern computing algorithms even allow an approximation of the original microscopic specimen to be constructed from an incomplete or garbled image file!

While I consider each of these 'principles' to have some basis in the relevant literature discussed previously, it is important, however, to emphasize that the list is incomplete and that these are actually hypotheses which may or may not be supported by future science education research. They are like the first map of a territory — somewhat helpful, but sure to require modification.

Specific Pedagogical Suggestions Based Upon Research

Distinguishing Between Magnification and Resolution

To simplify, some high-tech microscopes give biologists the opportunity to see their specimens enlarged at magnification powers several orders of magnitude higher than the light microscope offers. Yet, magnification alone is not enough. Magnification (image enlargement) without adequate resolution (the ability to distinguish two adjacent points on the image of the specimen) is not helpful to researchers since it merely produces big, fuzzy images. Each point on the specimen is represented by a corresponding disk on the magnified image; when a certain magnification is reached, the detail in the specimen becomes the same size as its disk on the image and further magnification will not improve the image quality (Healey, 1970). The naked eye cannot separate two dots that are less than 0.1 mm apart; the light microscope can separate two dots as little as 0.001 mm apart (see Bowlt, 1983, for a laboratory activity on measuring the resolution of a light microscope). Electron microscopes, because they use electron beams instead of light beams to form their images, not only offer high magnification power, but also provide high resolution (for example, resolving powers more than 100 times greater than the light microscope). It is vital to provide visual and diagrammatic examples of resolution and magnification for students to see before attempting to teach those students biological concepts using high-tech micrographs.

The Importance of Providing Both Examples and Non-Examples

Novak (1977, p. 139) considers the poor selection of examples by the teacher to be a major factor in explaining inadequate concept learning and concomitant lack of transfer to new problem situations. Research has shown that students who are given examples of an abstract category when the category is first introduced are able to apply the concept better to related but different learning tasks in later life (Cheng, Holyoak, Nisbett and Oliver, 1986; Gick and Holyoak, 1983). Moran and Rowley (1987) have found that first-year medical students learned histology better when they used 'correlative microscopy' — comparing pairs of LM and TEM micrographs of the same tissue taken at the same magnification. Meaningful learning theory suggests that practice in sorting out the differences in meaning between two related concepts (integrative reconciliation) is important if we want the concepts to be learned meaningfully (Novak and Gowin, 1984, p. 97). Hence I devised and used the following novel experiential activity, intended to provide both examples and non-examples of magnification and resolution in the biology classroom. The activity rests on the belief that careful comparison leads to understanding.

A Simple Activity to Compare Examples of Magnification and Resolution

Students are given a high-contrast photograph which was cut out from a newspaper and the following instruments: (i) a photographer's reducing glass (source: Edmund Scientific Company, Barrington, NJ); (ii) an assortment of magnifying glasses of various magnifying powers; and (iii) a laboratory ruler. All of the magnifiers have labeled powers. In keeping with the recommendation of Weakley (1981) that novices should write down their own microscopic observations in rich detail ('It is surprising how the act of putting what one sees into words makes one see more', p. 122), students are asked to record all that they notice within the photograph in great detail. The following procedure is then explained and the activity begins:

1 Examine the screened newspaper photograph with the 8 cm diameter, 2X magnifying glass, raising it gradually over the photograph until you see the largest, sharpest image it can give you, and then note and record the qualities of the resulting image.
2 Examine the same photograph with the 8 cm diameter *reducing glass* at a working (lens-to-specimen) distance of 10 cm (where its power is -2X). Compare the first instrument's image qualities with this one's. Record results.
3 Continue to examine the same photograph with the reducing glass at increasingly greater working distances until you reach its limit. How is the image you see changing in quality? Record your observations.
4 Using the same photograph as object, contrast the quality of the smallest photographic image you saw (using the reducing glass) with the qualities of the photographic images produced by a 10X and a 20X hand lens, respectively. Record your observations.
5 Questions to stimulate writing: What have you observed about the relationship between magnification and resolution when you consider the size and the degree of sharpness of the images you saw? What is the opposite of magnification? What is the opposite of resolution? How did the working distance (the distance from the photograph to the magnifier) and the field of view change as magnifier power increased? If your eye has a minimum working distance of ~25.5 cm, how does your hand look when it's held 2.5 cm away from your eye in comparison to 25.5 cm? Explain this basic event using the concepts of *magnification* and *resolution*.

Using the Hand-Held Monitor Microscope to Help Students Enter the Living Microworld

Imagine a 2 meter cable with a flashlight-size probe on one end. The other end of the cable is plugged into a small metal box that is, in turn, plugged into a television set. The box contains a halogen illuminator and its light is transmitted through the cable to the probe. A lucite collar on the probe holds any specimen at the right distance from the lens of the tiny color video camera inside the probe. Imagine a microscope that both an elementary teacher and a college professor can use in class without stopping to adjust illumination and focus. Imagine a living ant, 50 cm long, crawling across a television screen.

In 1991, while reading an American industrial engineering publication, I noticed an advertisement for a hand-held, ring-lighted, video microscope which featured a probe the size of a small flashlight which was designed to allow electronic repairpersons to reach inside dimly lit electronic hardware to inspect the circuit boards. Company representatives for the microscope manufacturer (Keyence) had never even considered that their new industrial inspection tool might serve an important pedagogical function. The potential of this lightweight, portable, easy-to-use tool for engaging students in microscopy was clear. Electron microscopist Cindy Henk of Louisiana State University was instrumental in helping me test and then introduce this new tool to the scientific community.

After individual use by hundreds of students and teachers from kindergarten through graduate school, and from surveys of these users, the following pedagogical conclusions may be drawn about this instrument:

1 This tool, which requires no focusing adjustment by the user, has its own light source for proper illumination of the specimen via fiberoptic cable, and whose probe can be held in even a small child's hand, allows students to enter and begin to explore the living microworld by first enlarging the external features of their own person (for example, examining their own skin and scars; finger prints, nails, and cuticles; hair, eyebrows, and whiskers; pierced ears; and the intricate alignment of cotton fibers and threads that make up the clothing they are wearing) and then exploring common living microworlds such as the parts of a flower, the appendages of a honeybee, ants crawling about a terrarium, mold growing on bread, and euglenoids in pond water. (Since the monitor microscope's camera is fitted with several lenses of varying magnification, students can also see what is gained and what is lost by switching to a higher magnification; Wandersee, Henk, Cummins and St. Julien, 1992).

2 By linking the monitor microscope to a color television monitor with a built-in videocassette recorder (VCR), students and teachers can make a permanent record of their own observations while using the microscope. The student's own recorded tape can even be taken home for viewing with family members. Teachers can use the very images that students in the class produced at the beginning of the school year for teaching specific biological content lessons at a later date.

3 Although such video microscopes may be too expensive for a single school to purchase (~$10,000 at this writing), K-12 school systems can often afford to make such purchases, ultimately reaching every student in the system by rotating the instrument through all of their life science and biology classes. Manufacturers such as the Keyence Corporation of America, the Moritex Corporation, and the HiROX Company now offer similar monitor microscopes with hand-held probes which we have examined and found suitable for biology teaching. (As an aside, the life scientists at our university were so impressed by the quality of the images and the ease of use of these instruments that some are now ordering them for conducting their own research — a rare instance of science education influencing science research practice!)

4 We found that the monitor microscope was ideal for teaching basic principles of micrometry. After we magnified a transparent millimeter ruler

Figure 12.1: Advantages of the monitor microscope

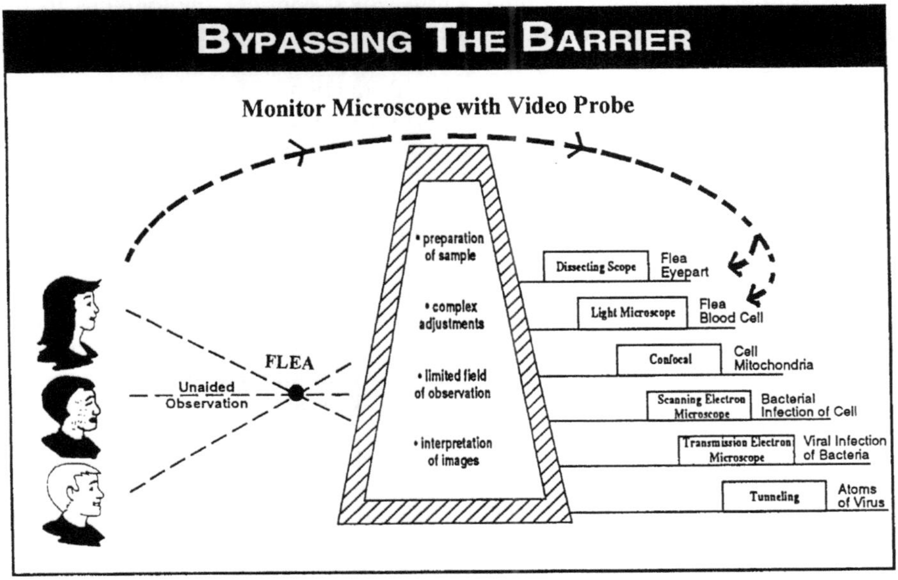

Note: This figure was adapted from a 1992 microscopy research grant proposal by H. Silverman, M. Sundberg, M. Socolofsky, J. Wandersee, C. Henk, C. Cummins, K. Thompson, R. Good, S. Pirkle, and J. Lynn – all of Louisiana State University, Baton Rouge, LA.

and observed the width of the field (television monitor), we asked students to image and measure various biological specimens. Students were surprised to find that when we changed magnification powers, the magnified ruler indicated that the dimensions of the field of view had changed quite dramatically. The ability to see the ruler and its image at the same time seemed to help students understand micrometric principles that had been more difficult for them to grasp using the light microscope. Because this microscope is stageless and can accept larger, three-dimensional specimens as measurement problems, the students could use specimens for which they knew the actual size from everyday experience. Figure 12.1 summarizes the major advantages of this new instrument and shows the range of images it can generate.

Using Small Multiple Graphics and Textual Analogy to Teach with High-tech Micrographs

Graphic theorist Edward R. Tufte (1983) invented the term *small multiple* to indicate a data-based graphic that is repeated on the same page so that 'once viewers understand the design of one slice, they have immediate access to the data in all the other slices' (p. 42). The essence of the small multiple is comparison of multiple data sets over time. Tufte would consider photographs to be graphics with a high data-ink ratio, since each grain in the film emulsion that produced the micrograph print is a light intensity data point.

In teaching the prokaryote/eukaryote distinction, we used a new graphic design involving 'small multiple vertical arrays of TEM micrographs flanking a specially written narrative text' as a replacement for a traditional biology text's 'expository text accompanied by two captioned protypical examples'. Not only was the former more effective in teaching the targeted concepts and principles, it also increased the amount of time introductory college biology students spent viewing the high-tech images accompanying biological text (Wandersee, 1992a). Figure 12.2 shows the mock textbook page which we created for this study. Note the way in which the text attempts to involve the reader in image analysis and how friendly the text is (for example word origins, town-city metaphor) and how the range of examples of each type of cell (prokaryote, eukaryote) allows the viewer to see the pattern across instances — a key factor in concept formation. The entire page is designed for visual comparison, guided perception, and making biological meaning of the critical differences once the reader-viewer sees those differences (Wandersee, 1992b). Note, also, that the comparison is set up to run from presence to absence of a nucleus, beginning with eukaryotes (in reverse of the order in which the distinction is usually taught). This was done because visual processing research indicates that it is easier to find something that is present in a graphic than to notice that something is missing (Hearst, 1991). Although the control group used a well-respected introductory college biology text's page introducing the same concepts, we found that its protypical micrographs outside the textstream were less helpful to students than the small-multiple micrograph sets closely integrated with our text. Measures of actual learning and self-reported learning preferences both clearly favored this new way of using micrographs, which is consistent with learning theory and demonstrates the power of a more narrative-style text. Such a pedagogical approach ought now be considered by curriculum designers and textbook authors.

For example, in a test of visual perception (Wandersee, 1992a), only 37 per cent of the females and 42 per cent of the males in the control group could locate the nuclei in five new TEM micrographs of eukaryotic cells, however 46 per cent of the females and 69 per cent of the males in the experimental group could ($N = 237$). Only 39 per cent of the females and 23 per cent of the males taught with the traditional textbook approach were at least 80 per cent successful in applying the town-city cell analogy to five new micrographs, however, 47 per cent of the females and 81 per cent of the males in the experimental treatment group reached the 80 per cent performance level ($N = 237$). Current research is analyzing the performance differences attributable to the text and those attributable to the small-multiple, micrographic examples. Early results indicate that the use of small-multiple micrographs explains more of the variance than does the revised text.

If one wishes students to learn the prokaryote-eukaryote distinction in a meaningful way, the students must study many micrographs. Using these, the teacher must guide the learner to make biologically important, fundamental perceptual distinctions (Barlow, Blakemore and Weston-Smith, 1990) and to see patterns across examples (for example presence or absence of a nucleus, relative

Figure 12.2: Text created for teaching prokaryotic/eukaryotic distinction

Eukaryotic Cells

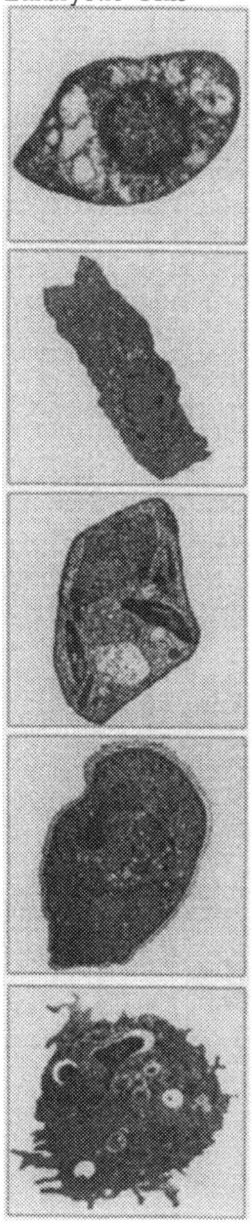

Prokaryotic Cells

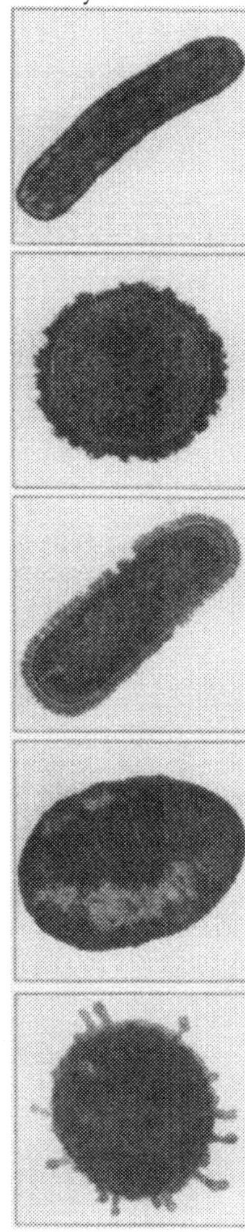

The Cell as a Community

The two groups of photographs on this page represent one of the most basic divisions of life -- cells with and cells without a nucleus. This division is so sharp that many biologists can recognize a cell as one type or another at a glance. Can you notice the differences?

All living organisms, including the simplest cell, can be regarded as communities. Like human communities, cells may vary in size and complexity. Each cell on the left can be thought of as a large city. Cities are divided into areas that perform specific functions like fire departments, power plants, water systems, and communication systems. In general, these cells are large and have distinct parts called organelles (little organs). Each type of organelle has a specific function for the cell as each area in the city has a specific function. You can see these organelles as distinct bodies in the pictures on the left. The largest of these organelles is the nucleus, which appears as a dark area with smaller dots within it. Can you find the nucleus in each picture on the left? The presence of this nucleus gives this type of cell its name, eukaryotic (*eu* = true and *karyon* = kernel [nucleus] in Greek).

Each cell on the right can be thought of as a small town. In small towns, many people have more than one job in the community and there is less division of labor into specific areas. In general, these cells are small, simple, and show fewer distinct bodies within the cell. Compare the cells on the left with those on the right to see if you can detect this difference. The absence of the nucleus gives this type of cell its name, prokaryotic (*pro* = before and *karyon* = kernel [nucleus] in Greek).

From the differences visible in these electron micrographs flow much of our understanding of the evolutionary history of life and some of the evidence for the 5 Kingdoms into which we divide all life. Prokaryotic cells are found only in Kingdom Monera while eukaryotic cells are found in the four remaining kingdoms. To review, how does the story about cities and towns relate to one of the most basic divisions of life?

size, level of microstructural detail [compartmentalization], etc.) in simple descriptive fashion — without the scientific labels. Once the learner can find these patterns, he or she can construct meaning by learning the scientific labels for those patterns (concepts) and by learning critical relationships between such concepts. 'Guided perception, concept formation, establishment of relations among and between concepts, development of constructs' — all of these lead to key cytological, systematic, ecological, and evolutionary principles based upon the prokaryote-eukaryote distinction (Wandersee, 1992a, p. 15).

While it is quicker to simply define *prokaryote* and *eukaryote* in words and then tell a student these principles, without an experiential base founded on 'basic concept-building activities' and interpretation of actual micrographs, the student is much more likely to employ rote mode learning strategies. No student should be experientially shortchanged; even the visually handicapped and blind can now use tactile micrographs printed with heat-polymerisable ink! While the temptation in teaching always is to rush, to try to create instant experts, it must be steadfastly avoided. Only students who know how to interpret the very micrographic evidence that scientists actually used to formulate those principles will be able to learn them meaningfully.

Learning *how* we know what we know in biology (interpretation of evidence) is as important as learning *what* we know (biological content knowledge). This chapter has suggested some guidelines, tools, and techniques that could become part of a 'construction kit' which teachers can use with their students in making biological meaning from the images of high-tech microscopy. However, to say that it describes a complete pedagogy of microscopy would be to boast *ex pede Hercules* (to measure Hercules from the size of his foot)! More research is needed in this domain if twenty first century biology students are to become biologically literate.

Acknowledgments

The author wishes to thank Marion Sokolofsky, Cindy Henk, Catherine Cummins, John St. Julien, Becky Demler, Sharon Matthews, and John Lynn for their intellectual contributions to the NSF/LaSER-funded *Exploring Microstructures* project on which parts of this chapter are based. Of equal importance and also deserving of thanks are the biology students and faculty at Delgado Community College in New Orleans, LA, all talented collaborators with the research team. The author, however, is solely responsible for any errors of commission or omission which may appear in this chapter.

References

Ausubel, D.P., Novak, J.D. and Hanesian, H. (1978). *Educational Psychology: A Cognitive View*. New York: Holt, Rinehart and Winston.

Barlow, H., Blakemore, C and Weston-Smith, M. (Eds.). (1990). *Images and Understanding*. Melbourne, Australia: Cambridge University Press.

Bowlt, C. (1983). Measurement of the resolution of the optical microscope. *Physics Education*, *18*(4), pp. 78–9.

Burgess, J., Marten, M. and Taylor, R. (1987). *Microcosmos*. New York: Cambridge.

CBE Scientific Illustration Committee. (1988). *Illustrating Science: Standards for Publication*. Bethesda, MD: Council of Biology Editors.

Chan, C.K.K., Burtis, P.J., Scardamalia, M. and Bereiter, C. (1992). Constructive activity in learning from text. *American Educational Research Journal*, *29*, pp. 97–118.

Cheng, P.W., Holyoak, K.J., Nisbett, R.E. and Oliver, L.M. (1986). Pragmatic versus syntactic approaches to deductive reasoning. *Cognitive Psychology, 18*, pp. 293–328.

Crockett, L.J. (1978). A dutch draper, his microscope, and his world of animalcules. *Science Digest*, *83*, pp. 36–41.

Dobell, C. (1932). *Antony van Leeuwenhoek and his "Little Animals"*. New York: Dover Publications.

Darius, J. (1984). *Beyond Vision*. Oxford: Oxford University Press.

Fisher, K.M. (1990). Semantic networking: The new kid on the block. *Journal of Research in Science Teaching*, *27*, pp. 1001–18.

Ford, B.J. (1985). *Single Lens: The Story of the Simple Microscope*. New York: Harper and Row.

Frazier, D.W. (1991). *'Directed and spontaneous transfer of college developmental reading students' textmarking strategies'*. Unpublished doctoral dissertation, Louisiana State University, Baton Rouge, LA.

Gick, M.L. and Holyoak, K.J. (1983). Schema induction and analogical transfer. *Cognitive Psychology*, *15*, pp. 1–38.

Gowin, D.B. (1981). *Educating*. Ithaca, NY: Cornell University Press.

Gray, P. (Ed.). (1961). *The Encyclopedia of the Biological Sciences*. New York: Rinehold Publishing.

Hardin, G. and Bajema, C. (1978). *Biology: Its Principles and Implications* (3rd ed.). San Francisco, CA: W.H. Freeman.

Hart, M.H. (1978). *The 100: A Ranking of the Most Influential Persons in History*. New York: Hart Publishing.

Healey, P. (1970). *Microscopes and Microscopic Life*. New York: Bantam Books.

Hearst, E. (1991). Psychology and nothing. *American Scientist*, *79*, pp. 432–43.

Henk, C., Territo, B., and Wandersee, J.H. (1992, October). *Microbial neighborhoods* (TEM Imaging for instruction). Paper presented at the annual meeting of the Louisiana Society for Electron Microscopy, Southeastern University, Hammond, LA.

Hoz, R., Tomer, Y. and Tamir, P. (1990). The relations between disciplinary and pedagogical knowledge and the length of teaching experience of biology and geography teachers. *Journal of Research in Science Teaching*, *27*, pp. 973–85.

Matthews, M.R. (1992). History, philosophy, and science teaching: The present reapprochement. *Science and Education*, *1*, 11–47.

Mintzes, J.J., Trowbridge, J.E., Arnaudin, M.W. and Wandersee, J.H. (1991). *Children's biology: Studies on conceptual development in the life sciences*. In Glynn, S. Yeany, R.H. and Britton, B.K. (Eds.). *The Psychology of Learning Science*. Hillsdale, NJ: Lawrence Erlbaum Associates, pp. 179–202.

Moran, D.T. and Rowley, J.C. (1987). Biological specimen preparation for correlative light and electron microscopy. In Hayat, M.A. (Ed.). *Correlative Microscopy in Biology: Instrumentation and Methods*. London: Academic Press, pp. 1–22.

Nist, S.L. and Kirby, K. (1989). The textmarking patterns of college students. *Reading Psychology*, *10*, pp. 321–38.

Novak, J. D. (1977). *A Theory Education*. Ithaca, NY: Cornell University Press.

Novak, J.D. and Gowin, D.B. (1984). *Learning How to Learn*. New York: Cambridge University Press.

Shulman, L. (1987). Knowledge and teaching. *Harvard Educational Review*, *57*, pp. 1–22.

Shulman, L. (1992). *Pedagogical Content Knowledge. Usefully Wrong?* Paper presented at the annual meeting of the American Educational Research Association, San Francisco, CA; April.

Singer, C. (1950). *A History of Biology*. New York: Henry Schuman.

Tufte, E.R. (1983). *The Visual Display of Quantitative Information.* Graphics Press, Cheshire, CT.

Wandersee, J.H. (1980). *A Voice from the Past: Antony van Leeuwenhoek.* Paper presented at the national convention of the National Association of Biology Teachers, Boston, MA; October.

Wandersee, J.H. (1981). A letter from Leeuwenhoek. *The American Biology Teacher, 43*, pp. 450–1.

Wandersee, J.H. (1986). Estimating the size of 'little animals.' In Schoen, H.L. and Zweng, M. (Eds.). *1986 Yearbook: Estimation and Mental Computation.* Reston, VA: National Council of Teachers of Mathematics; pp. 220–2.

Wandersee, J.H. (1992a). *The Graphic Representation of Biological Knowledge: Integrating Words and Images.* Paper presented at the NATO Conference on the Structure and Acquisition of Biological Knowledge, University of Strathclyde, Glasgow, Scotland; June.

Wandersee, J.H. (1992b). *Using Small-multiple Graphics to Improve Biology Textbooks.* Paper presented at the 24th Annual Conference of the International Visual Literacy Association, University of Pittsburgh, PA; October.

Wandersee, J.H., Henk, C., Cummins, C.L. and St. Julien, J. (1992) *A New, High-tech Microscope that Magnifies Interest in Science.* Paper presented at the annual convention of the Louisiana Science Teachers Association, Lafayette, LA; October.

Weakley, B.S. (1981). *A Beginner's Handbook in Biological Transmission Electron Microscopy.* Edinburgh, Scotland: Churchill Livingstone.

Chapter 13

Year 9 Bodies

Anne Symons, Kate Brass and Susan Odgers

Rationale

Human biology has traditionally been taught in a very content-dependent way. Teaching strategies remain similar from early secondary to upper secondary classes and seldom take account of the different developmental levels of the students. Assessment is equally inappropriate, generally involving a considerable amount of rote learning, with 'good' science students being those who have good memories for facts. The result of this practice is ineffectual learning and a loss of confidence in many students when ability is judged in relation to test results. Senior students remember very little of the content they were taught in their junior years and real understanding of the important concepts is even less evident. The traditional approach, with its emphasis on compartmentalized knowledge, does not lead to a linking of new ideas with existing concepts or give an overview of the functioning of the whole organism.

A New Approach

This chapter describes the method adopted by the Science Department at Korowa Anglican Girls' School to introduce a more effective teaching and learning process. Not only was a less abstract approach to learning needed but it was also hoped to make studying biology a more participative experience for the students by giving them opportunities to take responsibility for their own learning. At the year 9 level there is no set syllabus in the State of Victoria, nor are there public examinations, so the opportunity existed to follow a worthwhile goal. It was decided to develop a unit of work along these lines using the human body as the context. The content to be covered included the size, shape, position and overall role of the most important organs of the body, and the function of the circulatory system in linking those organs was to be the major concept.

Getting Started

At the time of the initial development of this unit of work some teachers had been exposed to ideas coming from the Children's Science Group at Monash

Figure 13.1: Initial probe

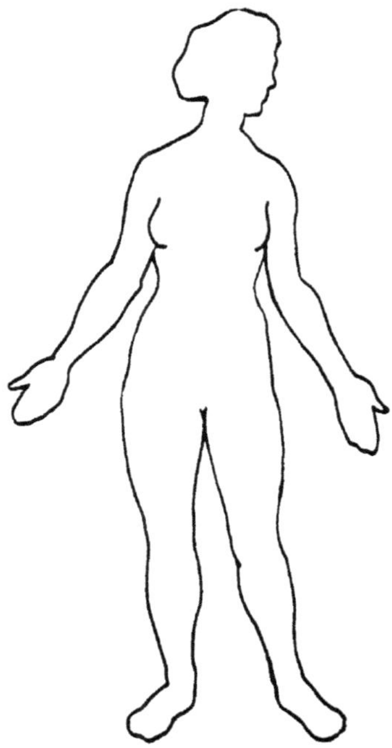

On the diagram above, mark in the size and position of the following organs: (i) heart; (ii) lungs; (iii) stomach; and (iv) kidneys

University. The significance of probing students for their prior understandings was quickly realized. A probe was developed to determine where the students thought the various organs were situated and their relative sizes (see figure 13.1). Invariably the drawings revealed inappropriate ideas and misconceptions (see figure 13.2). In addition to providing the teacher with useful information, the probe served to stimulate student interest in the topic and provided a benchmark by which they could assess their own learning at the end of the unit.

Working Together

A good way to manage this work seems to be to arrange the students into groups of two. Students of this age (13–14) do not necessarily have well-developed co-operative skills, so a lesson on how to work in a group can save frustration and friction later on. The advantages and disadvantages of group work are listed and strategies for overcoming problems are discussed. A typical difficulty arises when one student feels that she does more than her fair share of the task. An easy way of preventing this situation occurring is for the group members to set up contracts

Figure 13.2: Student responses

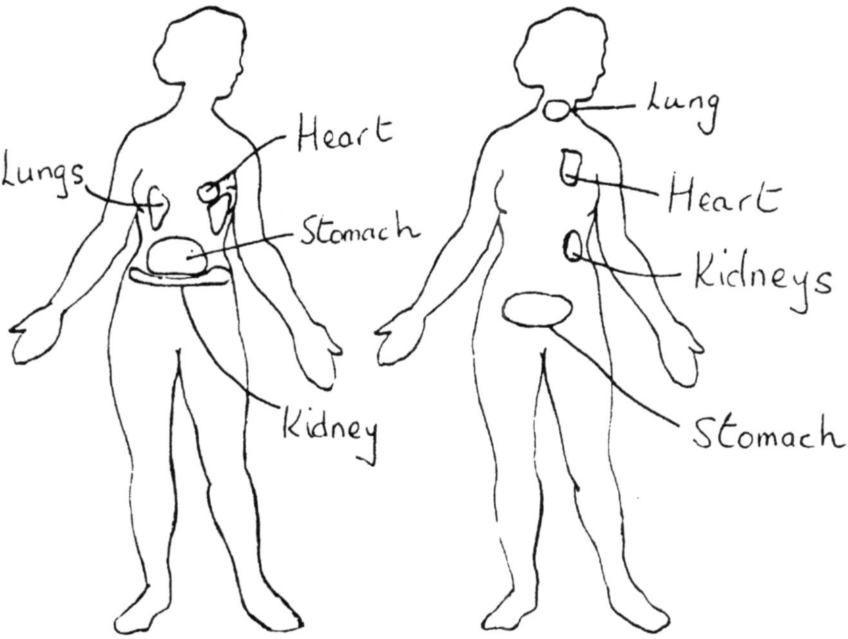

with each other which list the parts of the overall task to which each student is to be responsible.

A more difficult situation can arise when students who are selected for a particular group are unwilling to work together. They need help to understand that it is possible to learn from each other and to overcome the feeling that their understanding is something they have earned and is therefore not to be shared. Some students are very generous in the group-work situation; others want credit for what they do themselves and don't want to be associated with work they consider to be inferior. Long-term group work such as this allows students the time to discover their own strengths and also those of other students, some of whom they may have tended to dismiss as having little to offer.

Stages in Teaching and Learning Strategy

The next step is for the students to make a life-sized outline of the body on a piece of butcher' paper. They are given a list of the organs which must be included in their bodies — heart, lungs, liver, small intestine, stomach and kidneys — and told that they must also include a leg muscle. Their task is to place drawings of these components on the body and eventually link them with blood vessels so their model 'works'. The students have access to a range of books in the laboratory but quickly realize that the diagrams in most texts are misleading. A scale is rarely included and the positioning and shape of the organs is often inaccurate. To get the students started they are told that the heart is approximately the same size as the fist. This gives them a reference point from which they can work out a scale from more abstract diagrams. When they see a liver in a diagram they can then work out how big it is relative to the heart in that diagram, and get an approximation of its actual size. A lot of learning is involved in the selection of appropriate models and the development of dissatisfaction with models that are too simple is of great value to students. Eventually students will pick out the most sophisticated model that they understand and can use. Something which many students find very frustrating at first is the idea that there is often no single 'right' answer. They expect there to be a correct colour and a perfect shape for each organ. They should be encouraged to realize that what they are working on is their interpretation of the many different models they see in books and on videotapes.

Students do all the necessary research themselves, sometimes guided by leading questions from their teacher. Choosing the right moment to get involved is difficult. However, it is important on one hand to avoid letting the students feel discouraged and on the other hand to let them feel that they are capable of solving the problem at hand. The initial questions are likely to be along the lines of 'What colour is the liver?' or 'Is the stomach under the lungs?' Perhaps one group can be helped by referring them to their dissection of a sheep's pluck (heart, lungs, and liver) whilst another group may be encouraged to think about the problems presented by a two-dimensional representation of a three-dimensional object. As their work proceeds, the questions asked become increasingly complex. Students who ask questions such as 'Why is the liver so big?' can be told to find out what they can about the various functions of the liver. They will then be able to explain to other students some of the functions performed by the liver and will become the 'class experts' on the liver. This approach to students' questions has a valuable spin-off in the form of peer tutoring, which is a great confidence booster for the tutors and frees the teacher to deal with new problems.

Different students take differing approaches when placing the organs in the body. Some draw their organs directly on the body outline. Others draw them on separate pieces of paper and then stick them in position. Some groups are very particular about the details, so they create flaps and pop-ups in an attempt to show three dimensions. The greatest challenge of the design stage comes with the linking of the heart and other organs by blood vessels. Once they think that they have completed the task of connecting up all the organs, the students are given a set of counters that represent different kinds of molecules, in particular oxygen, carbon dioxide, glucose and urea. They have to be able to move these counters around the body from organ to organ, through the heart and the lungs, using the appropriate blood vessels. It is an unusual group that does not have major problems at this stage, particularly with the connections between the heart and lungs. Despite all the work they have put into their model it is not until they have to use it that most groups realize that most organ to organ movement involves the pulmonary circulation. It is vital to the learning process that the students are allowed to make mistakes in the construction of their bodies and that the teacher resists the urge to take control in the solution of their problems. They need to be encouraged to think first and then construct good questions to ask their peers or the teacher.

As a result of the method used to develop their models, there will be no two 'bodies' alike in any one class. There will be no 'correct' model. This idea is often most difficult to grasp for students who routinely achieve high grades. They seem to be less likely to take risks than those students who have 'less to lose' (i.e., whose previous record of success is not good). Small-group work often promotes a willingness to guess that would be unlikely from adolescents in a situation with the whole class as an audience. With encouragement, and an atmosphere that values all reasonable suggestions, even the most reluctant risk-taker will take a gamble. All students will eventually realize that the criteria for success have changed.

There is no point in using this sort of approach unless sufficient time is allocated to develop all the skills, understanding and insights that are possible.

Student Responsibility

It is not very often that students have the opportunity to develop long-term time management skills. They are far more adept at planning a task that will take an hour or a week rather than ten weeks. This of course means that they will need some strategies to help them. They are given a planning sheet which lists the major elements of the task and contains a series of boxes they can tick off once a section is completed. They are also required to keep a log in their workbook with a day-by-day record of their achievement. Even if it contains only one sentence, their log is a way for students and the teacher to monitor progress. When students have to keep a log of their activities they are made responsible for the way they spend their time. This gives the teacher more time to talk to each group.

The Need for Variety

Interspersed with their work on their bodies the students will spend time on various other activities to provide variety and to make points that could otherwise be missed. Examples of these are:

1 Dissections: sheep's pluck (heart, lungs, liver, etc.); heart; kidney;
2 Videotapes: digestive, circulatory systems;
3 Newspaper article analysis:
   (i) structured questions on content;
   (ii) word meanings, glossary work;
   (iii) discussion questions on social/ethical aspects;
4 Discussion of the history of blood transfusions;
5 Practical work:
   (i) measurement of lung capacity;
   (ii) heart rate — response to exercise;
   (iii) listening to heart sounds.

Student Learning Outcomes

The following learning is expected as a result of this unit of work:

1 An appreciation of the size and position of the major body organs;
2 An understanding of the role of the circulatory system in linking the parts of the body;
3 An understanding of the circulatory system as a double circuit system;
4 The ability to work effectively with a partner;
5 The ability to ask good questions;
6 An understanding of the role of models in science.

The philosophy underpinning these learning outcomes is to allow students to make mistakes, thus ensuring that they modify their ideas. Discovering that their body does not 'work' and then having to fix the problem, increases the likelihood of successful learning.

Assessment

From the beginning of the unit, the format and content of the assessment is made clear to the class. Students are required to move coloured counters around their body to represent the movement of molecules in and out of the bloodstream. They can perform as many practice runs as they need and a great deal of peer tutoring goes on. They are not tested until they feel prepared. Even when testing has started and it becomes apparent that there is a significant problem, the teacher will withdraw to allow the group time to work things out, and testing will only resume when they are confident in their understanding. This is considered to be significantly different from traditional testing. The students have a great deal of time invested in their work and greater value should be placed in encouraging them to maximize their understanding than in finding out how well they perform under pressure.

Testing takes the pattern of an initial informal chat about the project, such as the difficulties encountered and how they were overcome. Most groups find great satisfaction in the way they have been able to overcome problems, so they have considerable confidence in their ability to use their model. The initial questions asked of the students may be as straightforward as 'Show where an oxygen molecule enters the bloodstream?' or 'What is the name of the major vein leading up to the right side of the heart?' As they relax more difficult tasks can be asked of the students, such as 'Trace the path of a glucose molecule from where it enters the blood to where it leaves the capillary in a leg muscle'. Finally, for students who are still not particularly challenged by the regular testing, questions such as 'Does all the urea, in blood leaving the heart, go down to the kidneys?' or 'What happens to urea in the blood that travels to the brain?' will make them do some thinking. As the teacher is working with only one group at a time, it is possible to keep probing their understanding until their limit is reached. Not all class members will reach the same point. This type of assessment allows for variation depending on the level of student understanding and all students can feel that they have progressed in their level of knowledge.

Students are asked to assess their own understanding, and to write down the points that surprised them most about the structure or function of their body. This highlights the changes in their perceptions. Their original probes are also returned to them, serving to remind them of their initial conceptions and make them aware of the learning process they have been through. They derive satisfaction in fixing up their probes so they are 'correct'. If this is done in a different colour they have a permanent reminder of a worthwhile learning experience.

Student Perceptions of the Unit of Work

After the final testing, the students are asked to assess both the value of the task to them and the things that surprised them most about their 'bodies'. Some of the comments made by individuals in the year nine classes in 1992 were as follows:

> 'I think that I will remember what I have learnt because it was so hard'.
> 'I think I will remember because you can picture the diagram and store that picture in your brain'.
> 'This was a better way of learning because it was far more interesting than having to listen to a boring lecture'.
> 'This way was more interesting that listening to someone tell me and kept my attention better than books'.
> 'What surprised me the most was the fact the glucose and urea had to be pumped back through the heart and lungs and heart again to get where it needs to go'.
> 'The main problem we had was understanding the direction the blood was flowing in the heart. We also came across the problem that we couldn't move the molecules around the body, because we didn't think the blood had to go through the heart and lungs again. To solve our problems, we went back to the books and redrew the heart, so that it was clearer and we could understand it. I think I understand the heart a lot better now, because we did it again'.

Reporting

The assessment of the students' achievement in the unit may be reported in many ways. A descriptive assessment may be given based on such elements as:

1 Level of understanding of structure and function of the circulatory system, for example: 'The major organs of the body were correctly connected with arteries and veins, however, there was some difficulty shown in the understanding of the circulation of the heart and lungs (pulmonary circulation)';
2 Ability to work well in a group;
3 Time management skills;
4 Research skills.

This approach to learning is not incompatible with more formal testing. In a school situation which demands grades, the unit could be either graded in sections or each part, (for example, the model, the student's understanding of the function of the model and peer assessment) could contribute to a global grade. At Korowa there is not an evaluation of the model itself, preference being given to determining the level of understanding shown by the student in using it.

Conclusions

When using these teaching strategies it is important to realize that not all the students' questions will be earth-shattering revelations or insights. Writings on educational research do not often reflect the fact that, as well as deeply significant utterances, teachers are continually dogged by the 'I haven't got a blue pen, what will I do?' type of question. A great deal of time will be taken up initially with a plethora of requests of a mundane nature. Eventually the benefits in encouraging this type of learning experience will become evident. In addition to constructing their models of the body, the students will learn a great deal about several body systems in an integrated way and will begin to link new information with their considerable understanding of these systems. Students learn to analyze information from a variety of often contradictory sources and have the opportunity to develop a more mature relationship with their teacher. Working together cooperatively and making decisions contributes greatly to their personal development. In addition the students gain an awareness of the fact that models are developed to illustrate and explain ideas, and that a model reflects the understanding of its creator. In doing so they gain an insight, not only into specific content, but into the nature of science in general.

Acknowledgments

Korowa Science Staff involved in this project were Dorothy Brunckhorst, Susan Odgers and Anne Symons.

They are grateful for the administration of Korowa Anglican Girls' School for encouraging educational innovation and to the students of year 9, Korowa Anglican Girls School.

Chapter 14

Learning and Teaching Energy

Reinders Duit and Peter Haeussler

In this chapter we will attempt to investigate the teaching and learning of energy from a constructivist perspective. Explicitly included are issues we like to highlight from Fensham's (1986) 'Science for all' position, i.e., the interests of both girls and boys and STS, the interplay of science, technology and society. We will provide brief surveys of research findings that appear to be of importance regarding the issue of 'content specific pedagogical knowledge' (Shulman, 1987) in the domain of energy.

The Energy Concept in Science

Four Basic Aspects of the Energy Concept

Consideration of content specific pedagogical knowledge in the domain of energy must begin with an analysis of the science energy concept. A thorough analysis (Duit, 1986) led to four basic aspects of the energy concept that may also be called the 'energy quadriga' because they are intimately interrelated:

* energy transformation;
* energy transport;
* energy conservation;
* energy degradation.

The four basic aspects indicate that the science energy concept, on the one hand, stands for constancy amidst change but that there is, on the other hand, a decline of energy value whenever a process is taking place.

Transformation and Conservation

These two aspects undoubtedly are at the centre of the science energy concept. The idea of constancy amidst change was the most important factor of the energy concept in the history of science (Hiebert, 1962; Elkana, 1974). In the early nineteenth century, the science research program was much influenced by the romanticist idea of intimate interrelation between the 'forces' of nature. Researchers investigated the change of electricity into magnetism (Oerstedt in 1820), magnetism into electricity (Faraday's discovery of induction in 1831), heat into electricity (Seebeck in 1821), electricity into heat (Peltier in 1834) and many other phenomena. The concept of energy may be viewed as the scientific way of expressing the idea of a unifying, overarching romanticist 'force of nature'. It is important to know that this idea is still at the centre of the contemporary science energy concept. When we speak of energy transformation today, we have to be aware that the changes are occurring at the phenomenological level. The very idea of energy is that the amount of energy does not change despite all changes at the phenomenological level. At the conceptual level, there is only a change of energy's manifestations that are usually called energy forms.

Transport

A consequence of the view of energy transformation and conservation which has just been described is that energy usually changes place, so that energy becomes manifest in different places while processes are taking place. Quite frequently, energy transport is viewed as a 'flow' of energy (see below).

Degradation

Whatever processes take place in closed systems the amount of energy does not change but the usefulness of energy inevitably declines. To put it more precisely, at the end of every process that occurs in a closed system the number of further processes that are possible decreases. To indicate this, we therefore speak about energy degradation. Viewed on a microscopic scale, there is a tendency towards even distribution of particles and energy — as far as it is in accordance with the boundary conditions of a system. It may be surprising to find degradation among the basic aspects of the energy concept because in physics this is the domain of the concept of entropy. There are three major reasons to include energy degradation in the basic aspects of the energy concept, all of which are mainly pedagogical. First, energy and entropy are closely related to one another in science. But up to now in science instruction, the main emphasis has been placed on energy and the conservation aspect. The degradation aspect is still often neglected. Therefore, in school science teaching a key science concept is missing. Secondly, the conservation idea will become understandable for students only if the degradation aspect is also given attention (Duit, 1983). In all processes that take place in reality, there is the previously mentioned interplay of conservation and degradation. Hence, real processes are only understandable for students if both aspects are used to explain them. Thirdly, the degradation aspect is a key issue in STS approaches, i.e., in attempts to make students familiar with the energy problems of society.

Different Energy Conceptualizations

The basic aspects which are briefly described above form the core of the energy concept. But this core may be embedded in different frameworks, called 'energy

Figure 14.1: On the abstract, mathematical energy concept

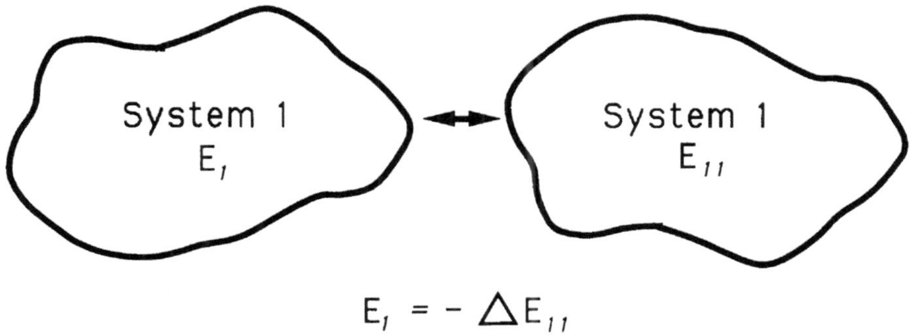

conceptualizations', that determine in which way the basic aspects may be interpreted. There are several different energy conceptualizations in the literature which usually have been developed to ease access to the abstract energy conceptualization as outlined in the following survey.

The Abstract Energy Concept

The 'genuine' physics energy concept is an abstract, mathematical one. There is a system (for example, system I in figure 14.1). In this system, quantities of state are defined, one of them being energy (EI). No matter what happens in this system, there is no change in the quantity of state energy. If the results of measurements are put into the appropriate energy formula, the same number is produced. If this number is not the outcome of a calculation, one has to search for transformations in the system that have not been regarded so far or else there has been an interaction with another system (system II). If the latter is the case, in one system the energy number increases whereas in the other system it decreases accordingly.

It is clear that the above mentioned basic aspects have a very abstract meaning within the sketched abstract energy concept. It becomes obvious here that energy is a book-keeping quantity. Energy conservation only means that the energy number of a system does not change, if there are no interactions with other systems, or that the energy number in the systems involved in interactions change in such a way that the sum of changes is zero. It is important to note here also that the physics energy concept may not be viewed as an entity that causes changes. The physicist Auerbach (1913) provided this metaphor to indicate the appropriate status of energy: in the big firm of nature, energy is only the book-keeping agency, the director entropy decides in which direction occurrences are to go.

The Idea of Energy Flow

Figure 14.1 highlights that the abstract physics energy concept is a quantity of state of a system. Therefore, energy can be viewed as being 'stored' in a system.

But the changes of energy numbers in interacting systems also allow a view of energy as being 'exchanged' between systems. In most practical uses, the exchange aspect is the more important one. Of course it matters how much energy is stored in a system but how much energy one may get out of a system depends on the changes the system undergoes (c.f. Carr and Kirkwood, 1988). Very often the exchange aspect is conceptualized by employing the idea of energy flow. This idea has become popular in science teaching. Energy is viewed as something that flows from one system to the other if there are interactions that result in the change of the energy numbers in the two interacting systems (as indicated in figure 14.1). This idea is helpful in easing access to the abstract energy concept because it provides a somewhat pictorial underpinning of the abstract change-of-numbers idea. The idea of energy conservation is very obvious within this conceptualization if the energy flow is viewed as the flow of an indestructible entity. Whereas energy transport is only a change of numbers, in systems in the above abstract conceptualization it may be viewed here quite pictorially as the flow of the entity energy.

Energy is the Ability to Perform Work

This conceptualization is still predominant in science textbooks even though it should be banned from science instruction. The reason is not that this conceptualization is wrong from the physics point of view as Lehrmann (1973) argued in an article that started a fruitful debate. There are several — mainly pedagogical — reasons to drop this conceptualization. First, the conceptualization may mislead students to think that energy is the cause of processes. Secondly, the implicit assumption that the concept of force is easier to understand than the concept of energy has been proven wrong. There are many studies now that show the concept of force is at least as difficult to understand as the concept of energy. Thirdly, the idea about the ability to perform work is in danger of reducing the energy concept to mechanics or at least to neglect that energy is the unifying concept that allows us to bring aspects of many areas of science together. Fourthly, the conceptualization leads to difficulties in understanding the second law of thermodynamics appropriately. If energy is seen as the ability to perform work and if this ability decreases in every process (which is one way of stating the second law of thermodynamics) then there is a certain contradiction to the principle of energy conservation.

The Energy Concept of Everyday Life

There are several studies of students' conceptions of energy both before the energy concept had been introduced in science instruction and after this had been done. The results show that conceptions both before and after energy teaching mainly reflect the use of energy in students' life-world domain; this is especially so where an energy concept is in use that differs from the science energy concept. There are also a few studies available on teachers' (Veiga, Costa Pereira and Maskill, 1989; Kruger, 1990) or teacher students' (Baird, Fensham, Gunstone and

White, 1987) conceptions of energy, which indicate that teachers and preservice teachers are often not in full command of the science energy concept and hold ideas that are similar to the ideas of their students.

Students' Conceptions of Energy

Students frequently conceive of energy as a universal kind of fuel. The fuel may be produced from certain sources like oil, gasoline, coal, sun, wind, or flowing water. Accordingly, for many students energy is something needed for machines in order to operate them. Energy is, as a whole, an industrial product which can be obtained from certain raw materials. Energy is also frequently affiliated with living beings, especially with humans, i.e., there is a certain anthropocentric view of energy.

A more detailed view of students' conceptions on energy may be provided by the set of 'alternative frameworks of energy' introduced by Watts (1983). We quote Lijnse's (1990) version of Watts' frameworks here:

> *Human centred energy*. This is an example of the frequent occurrence of anthropomorphic thinking, which with respect to energy especially finds expression in the idea of being energetic, having a lot of energy to be active, or having lost all energy and therefore feeling exhausted.
>
> *A depository model of energy*. This idea suggests a notion of energy as a source of power, as a cause of activity, and as such stored in objects that possess energy.
>
> *Energy as an ingredient or as a product*. In these cases energy is not so much considered to be a cause, but rather a result of, or a reaction to, something. Energy is found in food but only becomes active after you have eaten it. Or, energy is not stored in coal but manifested only after combustion.
>
> *Energy as an obvious activity*. In this notion, energy is identified with the presence of activity, particularly motion. Energy is motion.
>
> *Energy is functional*. In many situations energy is conceived to be a general kind of fuel, as something that is especially useful and has something to do with technology.
>
> *Energy as some kind of fluid*. Energy is then materialised as a kind of fluid that flows and may enter and/or leave something. (p. 574)

Whereas there are striking similarities between frequencies of students' ideas from different countries there are also interesting differences. Studies in the English language context have generally resulted in a considerable percentage of human centred ideas of energy and of associations with food (Solomon, 1983; Bliss and Ogborn, 1985; Gair and Stancliffe, 1988). Findings from Germany (Duit, 1981) have shown that framework 1 is very infrequent in this language context and that energy having something to do with food also appears to be an infrequent notion among German students. Trumper (1990) found that among Israeli high school students the percentage of the fuel idea of energy (framework 5) is much lower than in Lijnse's (1990) study with students of a comparable age in the Netherlands.

Use of Energy in Written Materials of the Life-World Domain

It was pointed out at the outset of this chapter that students' conceptions of energy reflect the use of energy in their life-world contexts. In fact, if one studies written material that is available in the life-world and that is written for the purpose of informing the public about energy issues, the way energy is used there is considerably different from the scientific use and appears to be nearer to the students' conceptions mentioned above. Lijnse's (1990) study supports this impression. Pamphlets, reports and newspaper articles written for the general public were analyzed with respect to the energy situations and concepts used. The conceptions of energy prevalent in the analyzed materials were very similar to the above students' conceptions. The central idea was often that we need energy for technical facilities and that energy is consumed when it is harnessed. Conservation of energy (as this principle is used in science) played no role in the analyzed literature.

The Energy Concept in Daily Life and in Science

If one employs the four basic aspects of the science energy concept (transfer, transformation, conservation and degradation) presented above to analyze the energy concept in daily life there are some striking dissimilarities. Most importantly, there is no conservation idea in a daily life context as there are present in science, but some key facets of energy degradation appear to be contained. This becomes apparent, for instance, if it is said that energy is used up in processes. But there is also a major difference between the energy degradation idea in science and in the daily life contexts. Whereas in science, energy degradation is closely affiliated with energy conservation, this is not the case for an everyday life context, simply because no conservation idea is present. It is thus not surprising that the research findings may be summarized that the daily life energy concept is nearer to degradation than to conservation (Duit, 1983). Ogborn (1986) has convincingly argued that the energy concept of daily life is quite close to the science concept of exergy. Exergy — roughly outlined — is the maximum energy that may be gained from processes in which a system interacts with its surroundings. Exergy, therefore, is the amount of useful energy in a system. Exergy in some way integrates the science concepts of energy and entropy, i.e., it is the energy gain from a system that is allowed by the second law of thermodynamics. Many uses of the word energy in daily life context become mainly 'right' if energy in the sentences referred to is replaced by exergy. This point of view has important consequences for drawing conclusions on how to introduce the energy concept in school science, as will be discussed below.

Results on Learning the Energy Concept During School Science Instruction

It is a well known 'sad' outcome of research on students' conceptions in whatever field of science and in whatever country that students, at best, gain minor facets of the science conceptions that instruction presented. It is, therefore, no surprise that rather limited outcomes of energy teaching also have been reported. The result of Lijnse's (1990) study that only 17 per cent of students' answers after physics instruction on energy fall into a category the researchers accepted as being 'physically correct descriptions' is quite typical. Some studies investigated the outcome of science instruction via interviews or questionnaires without observing the instruction given (for example, Duit, 1981; Kesidou and Duit, 1991). But there are also two seminal projects on developing new approaches to the energy concept on the basis of the constructivist view where instruction was carefully observed (Alanach *et al.*, 1987; Carr, Kirkwood, Newman and Birdwhistle, 1987). There are further studies in which constructivist ideas led to alternative methods of teaching energy (such as the study by Trumper, 1990 and 1991). Although the latest constructivist approaches appear to be more successful than traditional science instruction, the main deficiencies as outlined below appear to occur in these approaches as well.

Students Usually Do Not Learn the Basic Aspects of the Energy Concept

It has already been mentioned that students have severe difficulties learning the idea of energy conservation (see the findings in the study of Driver and Warrington, 1985, as another typical example). But students also have severe problems learning the basic aspect of energy transformation that is usually viewed as an easy idea. Duit (1981) compared energy conceptions of German students (in total about 300 students) in grade 6 before physics instruction and in grade 10 after four years of this instruction. The main changes to be observed were in the energy vocabulary. Students in grade 10 were able to give a lot more examples for energy forms than in grade 6. But that did not mean that the students learned the idea of energy transformation. A more recent interview study with thirty-five students in grade 10 (Kesidou and Duit, 1993) supported these findings and also revealed in which way students failed to learn the ideas of energy conservation. This principle is often mentioned by the students in the interviews as an important aspect of energy but it is not interpreted in the physics framework but in a students' 'cause effect' framework. Viewed in this framework energy is not lost in a process because an effect (for example, heating a body or lifting a weight) was caused — even if there is no energy any more at the end of the process.

Students Do Not Use the Energy Language as Taught in School When Explaining Processes

Another general finding is that students seldom use the scientific terms (here the science energy language) when they are asked to explain processes. If these processes are similar to those they are familiar with (and that have been discussed in class) they usually employ the scientific terms. But if the processes are a little less familiar, students tend to rely on their everyday terminology and not on the scientific terms.

Students Do Not Use the Energy Knowledge They Learned in Science Instruction in Real Life Situations

It is an important aim of teaching energy in school to make students able to understand occurrences in their life world domain by employing energy ideas taught in school. There are not many studies that suggest whether this is the case. But there are, of course, doubts that it will take place because students frequently do not even employ science knowledge in a science context as has been outlined above. Arzi (1988) reports an interview study on energy in chocolate and yogurt. She presented students with, for instance, a packet of chocolate labelled 'energy' and then asked questions on energy in food. Students had severe difficulties in applying school science knowledge of energy. In a study by Duit and Zelewski (1979), thirty-four students (about 15-years-old) who had completed an energy course one year previously were asked by the German teacher to write an essay on 'Much is to be heard about energy these days. Please explain what energy is and please explain the importance of energy for us'. Although aspects of energy supply had been given attention in the physics course the students had followed, and although the teaching unit on energy lasted about twenty lessons, most of the knowledge presented in the essays stemmed from mass media and not from what had been taught in physics. That was even true for explanations on 'what energy is'. Only a small number of students mentioned aspects they had learned in school, and even fewer did this in an appropriate manner.

Results of a Study on Long-term Retention of Energy Knowledge

Here we will summarize findings of a study on long term effects of physics instruction carried out at the Institute for Science Education at the University of Kiel, Germany, (Haeussler, Hoffmann and Rost, 1986; see also Haeussler, 1987a; the summary here mainly follows Duit, 1989). The study surveyed, among other topics, for over 800 adults in Germany the level of theoretical knowledge in the area of energy, attitudes towards alternative energy technologies, attitudes towards — as well as active commitment to — energy saving and actual energy saving behavior. Findings of this study shed some light on the relevance of teaching the physical energy concept for understanding energy problems in society and for energy saving behavior.

The level of theoretical knowledge on energy of the adults depended — and this is no surprise — on the amount of physics taught at school (and on some other variables such as the school type). But the amount of energy content was no better predictor than the total amount of physics taught. These results suggest that school physics appears to have a more general effect. That would mean that preparing students for understanding energy supply issues via physics energy knowledge does not necessarily demand devoting much time to energy teaching.

It is highly interesting that an adult's dealing with energy issues outside school had a much stronger influence on attitudes and energy saving behavior than school variables. In addition, there seemed to be no connections between a person's knowledge of energy, on the one hand, and attitudes towards alternative energy supply technologies and attitudes towards energy saving on the other. Among the variables on the type of physics learning, 'understanding and discovering the laws of physics' appeared to be of pivotal importance. There were positive connections, not only to the theoretical knowledge but also to most attitude variables. This finding suggests that the particular content taught in school appears to be of minor importance compared with guiding students to a positive leaning towards physics knowledge and inciting their interests in this field. The study, therefore, permits the interpretation that instruction which simply informs students of physics significance for society and everyday life without developing their competence to act and without aiming at students' attitudes may turn out to be relatively ineffective.

On Energy Knowledge and Energy Saving Behavior

It seems to be an everyday experience that knowledge about a field and actual behavior in it are not necessarily in harmony with each other. Another Institute for Science Education at the University of Kiel, Germany, (IPN) study (see Langeheine and Lehmann, 1986; the summary presented here mainly follows Duit, 1989) investigated relations between knowledge in the field of environmental protection and the actual behavior of people. Energy aspects were also contained in this study.

The level of knowledge in the field of environmental protection was mainly influenced by two variables, namely 'level of education in school' and 'search of people for topics of environmental issues in mass media'. The latter variable was considerably influenced by a variable that may be called 'experiences with nature (for example, living beings) through childhood and adolescence' and a little by the previously mentioned 'level of education'. Interestingly enough, the level of 'environmental knowledge' was not among the variables that mainly influenced actual 'environmental behavior'. The variables mainly related to public commitment to environmental issues as well as actual environmental behavior in daily life (for example, energy saving in daily life, collecting of raw materials such as glass) were the previously mentioned 'search of people for topics of environmental protection in mass media' and 'attitudes towards economy growth'. Also experiences in childhood with careful treatment of living beings (for example supported by the parents) played a certain role. Briefly put, the study points out that 'environmental knowledge' appears to be of less importance concerning 'environmental behavior' than a tendency to search for 'environmental topics' in mass media, attitudes towards economy growth and experiences with careful treatment of living beings in childhood and adolescence. There has been a wealth of studies carried out elsewhere that show similar results. In a meta-analysis of seventeen studies in the field of environmental protection Hines, Hungerford and Tomera (1987) found only small correlations between knowledge and actual behavior.

Gender Issues of Students' Interests in Physics Instruction

Gender issues have been given a great deal of attention in science education research over the past decade. Physics and also chemistry have proven to be the science domains that are greeted with least interest by girls. But also the interests of boys in these domains are not always satisfied. We will report briefly here a project investigating students' interests in German physics instruction of grades 5 to 10 (Haeussler, 1987b; Haeussler and Hoffmann, 1990). About 4000 students were involved. The design allowed conclusions on the development of interests from grade 5 to 10. We will summarize major findings of this study in the following:

(A) The interest on dealing with 'society issues' in physics instruction is generally high; the interest of girls in these issues is higher the older they are and the more they are personally involved.

(B) To connect the content taught with issues of daily life stimulates interest, but for girls only if they may rely on their own experiences.

(C) Physics that arouses emotions in general stimulates interest. Girls' feelings appear to be excited especially by natural phenomena. They are also, but not as much as boys interested in 'stunning' technical devices or developments.

(D) Discovery of physical laws or introduction of these laws as issues for their own sake is not greeted with interest. This is especially true when dealing with quantitative issues. The interest in quantitative physics rises if an application (at least a potential one) is included and the necessity and/or value of quantification is made understandable for the students. It is favorable for girls if among the domains of application the focus is not solely on stereotypical male domains of technology but applications in medicine, in environmental protection concerns or issues of their own body are also regarded.

Aims of Teaching and Learning Energy in School

As has already been mentioned, there are two major focuses if teaching energy in school is to be justified. These were also stressed in a Delphi-study regarding desirable physics instruction (Haeussler, Frey, Hoffmann, Rost and Spada, 1988).

1 The relevance of energy in science, where it is claimed that energy is a key science concept that also allows access to key science processes.
2 The relevance of energy to understand occurrences in students' life-worlds, most notably to understand energy issues in society (STS issues) and to act in energy-concerned ways.

The Energy Quadriga and the Major Aims of Teaching Energy

The aims of teaching energy described above demand giving attention to all basic aspects of the energy quadriga, namely energy transport, transformation, conservation, and degradation. From a scientific perspective, the intimate interplay of all the basic aspects (including degradation) is necessary in order to make the other aspects, especially energy conservation, understandable. A life-world perspective highlights degradation because this is the crucial aspect in matters of energy supply and efforts of saving energy.

Issues of STS and Interests Integrated

It is a happy coincidence that orientation of science instruction to issues of society meets the interests of students, most notably of both girls and boys. In the study on interests in physics presented above, only items that fall into a STS orientation displayed a growing interest from grades 5 to 10 and also showed about the same interests for boys and girls in grade 10. Studies on attitudes towards energy supply issues resulted in similar conclusions. Kuhn (1979) found a slightly higher awareness of the need of energy conservation (in the sense of 'saving' energy) among girls whereas boys showed a little more interest in developing new energy resources. Lawrenz and Dantchik (1985) who used the same 'energy opinionnaire' as Kuhn (1979) found a growing interest in 'saving energy issues' when students become older. They also found gender differences of the following kind. Older girls 'were more in favor of conservation than boys, especially in the context of providing for others. Also the high school girls were overwhelmingly more in favor of governmental control than boys' (Lawrenz and Dantchik, 1985, p. 188).

On Chances to Set the Central Aims of Energy Teaching into Practice

The research results reviewed in the preceding sections of this chapter give some important hints of whether the aims of energy teaching described above are only dreams or whether there is a realistic chance of putting them into practice. Unfortunately, the results appear to support the 'dream side' because they show that the results of energy teaching (as also of teaching other science topics!) are far from the outcomes expected by curriculum developers and others who set aims for science instruction. We will briefly summarize the evidence given above.

Understanding the Science Energy Concept?

Students in school energy teaching usually do not learn the basic ideas, the elements of the energy concept, They mainly hold onto their life-world conceptions of energy. At best, some issues of the science concept are loosely integrated into these conceptions.

Employing School Science Energy Knowledge in STS Concerns?

The issue of STS may be divided into two aspects. The first claims that students should be able to understand issues of energy supply in society by employing knowledge gained in school. The second regards the impact of energy teaching energy-concerned behavior in daily life. Where understanding energy issues in society is concerned, the results show that energy knowledge gained in school is often not even used in school science contexts. Use of this knowledge in out-of-school domains does not appear to be very likely. Energy-concerned behavior does not appear to be directly influenced by energy knowledge but by more general factors that have been mentioned above.

Energy Knowledge to be Employed in Daily Life Concerns is Not Only a Matter of School Energy Teaching

The aims of energy teaching that fall into the STS orientation may not only be addressed by teaching STS oriented energy courses. It may even be much better to provide a climate that incites students' interests in science and supports students' confidence that science has to say something to them in a much more general way. In other words, there are no direct relationships between the amount of energy teaching and the amount of understanding energy issues as well as energy concerned behavior later in life.

Constructivist Issues — Bridging Between the Energy Concept of the Life-World Domain and Science

Here we consider how it may be possible to guide students from their everyday conceptions to basic aspects of the science energy concept despite the fact that there are substantial dissimilarities between the life-world and science conceptions that hamper understanding the latter.

Conceptual Change Approaches

The path from students' pre-instructional conceptions to the science conceptions may be viewed either as a continuous or discontinuous change. There are several terms in use to indicate this, such as weak and strong restructuring (Carey, 1986) or evolutionary and revolutionary change (for example, Nussbaum, 1983) alluding to Kuhn's (1970) views on paradigm shift. Hence, constructivist approaches fall into two categories. The 'evolutionary' oriented approaches start from aspects of students' conceptions that are mainly in accordance with the science conceptions to be learned. From there, a chiefly continuous passage is used. The 'revolutionary' approaches deliberately point to the dissimilarities. Accordingly, cognitive conflict is a typical strategy in these approaches (Scott, Asoko and Driver, 1992). It is important to note that from a constructivist perspective it is not assumed that at the end of a learning process the 'old' ideas (the students' conceptions) have been erased and replaced by the science conceptions as traditional approaches often appear to assume. The research results support the constructivist point of view that the replacement idea is not adequate because students still use their old conceptions in situations where they have been proven to be rather successful.

From Students' Everyday Conceptions of Energy Towards the Science Conceptions

The way we want to bridge between students' energy conceptions and the scientific ones is determined by the aims of instruction. We argue deliberately in favor of a school science energy teaching that does not only provide an insight into basic concepts, principles and processes of science. We think that the main focus should be the understanding of energy issues in everyday life and in society as well as energy concerned behavior. If this is the focus, it has far reaching consequences for the conceptual change we have to arrange. We have to reject the idea of replacing students' conceptions by scientific ones. If we were to try (it would most probably not work anyway), students would learn a conception that is at least partly contradictory to the life-world energy conceptions. Hence, it would not be very useful in life-world contexts.

We, therefore, think that an evolutionary approach has to be designed that starts from facets of the life-world energy concept that are already in accordance with the science concept. This appears to be possible. Regarding Ogborn's (1986) previously mentioned view that the use of energy in daily life is, as a whole, correct from the scientific point of view if one interprets it as 'available energy' (more precisely as exergy), the everyday use of energy has to be reinterpreted from the point of view of the science energy concept. If, for instance, it is said that energy is consumed in a certain process, this has to be reinterpreted by employing the above energy quadriga: energy is transformed in that process from one form to certain others and may change the place of manifestation (is transported) but the amount of energy does not change (conservation) whereas the value decreases (degradation). In short, we aim at a strategy to nullify the contradictions between the life-world and the science energy concept by interpreting the life-world concept in terms of the energy quadriga.

The sketched program of developing the life-world energy concept demands for a conceptualization of energy that is near to the framework in which energy is embedded in life-world and that is also justifiable from the science point of view. We think that the following conceptualization may meet these conditions. Energy is viewed as a flow of a quasi-material something, as a kind of stuff that flows and that changes manifestations during processes. This quasi-material something may also be interpreted as a general kind of fuel. It appears to be obvious that it is near to the life-world energy concept. We think it is also acceptable from the science point of view although we do not deny that there are certain dangers that this view may invite misinterpretations. Where the basic aspect of energy conservation is concerned, that is not contained in the life-world energy concept, the quasi-material view appears to be quite helpful in strengthening these ideas, as preliminary findings of Kesidou and Duit (1991) have shown.

Towards New Approaches of Teaching Energy in School

The purpose of the preceding sections of this chapter was to develop major aspects of 'energy specific pedagogy' that point to issues that should be addressed when developing teaching and learning sequences in the domain of energy. In the following concluding remarks we want to outline very briefly in which way the ideas presented above may be put into practice.

We first want to refer to proposals of teaching energy by Lijnse (1990) that aim at combining issues of STS and constructivism. They fit our ideas to a considerable extent. Lijnse's energy sequence comprises three levels. The first is called 'ground level'. Here the focus is on relevance of energy knowledge for the adequate understanding of energy issues and adequate acting in the life-world domain. The main issue is linking up with students' ideas, i.e. starting from students' ways of talking and thinking with respect to energy in life-world situations. The second level is called 'quantitative life-world level'. Also here, students' life-world conceptions mainly stay intact. Energy is, for instance, viewed as something material which gets lost, to a certain extent, in every process. Quantification allows to calculate this loss. These two levels are viewed by Lijnse as mainly sufficient to address issues of energy concern in the life-world domain. The third level ('theoretical') focuses on introducing students to the 'real' science concept of energy. Whereas the change from the first to the second level is viewed as continuous, the change from the second to the third is seen as discontinuous.

Although we mainly agree with Lijnse's energy sequence we think that the approach we sketched in the previous section in some way avoids discontinuous changes. It is our idea to underpin, and by that to reinterpret, life-world energy conceptions by employing the basic ideas of the energy quadriga from the outset. Therefore, there is a continuous development of students' energy conceptions towards the science conception although major aspects of the life-world view stay intact. We propose to start teaching energy with about 12 to 13-year-old students. First, there is a focus on energy related phenomena and uses of the word energy in situations with which students are familiar. To address specifically girls' interests, students' self-made experiences should be given particular attention, as well as experiences that may emotionally move students' feelings and that take issues about their own body (such as energy in food) into consideration. STS comes into play only later when students' awareness and interest in these issues develop. Where quantification is concerned, it should only be introduced if it is possible to convince students that a quantitative dealing with energy (for example, calculations of energy gains or losses) provides interesting and valuable insights. In general, we think that our approach may be suited to address some of the learning deficiencies outlined above. Although the way we propose to proceed is very much influenced by energy life-world issues, the resulting energy concepts we aim at also appear to be acceptable from a position that is influenced by the relevance of energy in science.

References

ALANACH, D., ANDERSON, M., BROOK, A., DAVIDSON, J., DOW, A., DRIVER, R., GATER, S., HUGHES, D., KENT, D., POWELL, D., PRICE, B. and WELLS, P. (1987). *Approaches to Teaching Energy*. Leeds: Centre for Studies in Science and Mathematics Education: University of Leeds.

ARZI, H.J. (1988). *On energy in chocolate and yogurt, or: on the applicability of school science concepts to real life*. Paper presented at the annual meeting of the American Educational Research Association in New Orleans, LA; April.

AUERBACH, F. (1913). *Die Weltherrin und ihr Schatten*. Jena: G. Fischer.

BAIRD, J., FENSHAM, P., GUNSTONE, R. and WHITE, R. (1987). Individual development during teacher training. *Research in Science Education*, *17*, pp. 182–91.

BLISS, J. and OGBORN, J. (1985). Children's choices of uses of energy. *European Journal of Science Education*, *7*, pp. 195–203.

CAREY, S. (1986). Cognitive science and science education. *American Psychologist*, *10*, pp. 1123–30.

CARR, M. and KIRKWOOD, V. (1988). Teaching and learning about energy in New Zealand secondary school junior science classrooms. *Physics Education*, *23*, pp. 86–91.

CARR, M., KIRKWOOD, V., NEWMAN, B. and BIRDWHISTLE, R. (1987). Energy in three New Zealand secondary school junior science classrooms. *Research in Science Education*, *17*, pp. 117–28.

DRIVER, R. and WARRINGTON, L. (1985). Students' use of the principle of energy conservation in problem situations. *Physics Education*, *20*, pp. 171–6.

DUIT, R. (1981). Students' notions about the energy concept — before and after physics instruction. In JUNG, W., PFUNDT, H. and VON RHOENECK, C. (Eds.). *Proceedings of the International Workshop on 'Problems Concerning Students' Representation of Physics and Chemistry Knowledge'*. Ludwigsburg: Paedagogische Hochschule, pp. 268–319.

DUIT, R. (1983). Is the second law of thermodynamics easier to understand than the first law? In MARX, G. (Ed.). *Entropy in the School. Proceedings of the 6th Danube Seminar on Physics Education*. Budapest: Roland Eoetvoes Physical Society, pp. 87–97.

DUIT, R. (1986). *Der Energiebegriff im Physikunterricht*. Kiel: IPN.

DUIT, R. (1989). Science education and the quality of life — the case of teaching energy in German schools. In HONEYMAN, B. (Ed.). *Science Education and the Quality of Life. Proceedings of the ICASE World Conference: CONASTA 37*. Canberra: Australian Science Teachers Association, pp. 155–9.

DUIT, R. and ZELEWSKI, H.D. VON (1979). Ohne Energie ist es duester in unserem Leben. *Naturwissenschaften im Unterricht — Physik/Chemie*, *27*, pp. 161–64.

ELKANA, Y. (1974). *Discovery of the Conservation of Energy*. London: Hutchinson Educational.

FENSHAM, P.J. (1986). 'Science for all'. *Educational Leadership*, *44*, pp. 18–23.

GAIR, J. and STANCLIFFE, D.T. (1988). Talking about toys: an investigation of children's ideas about force and energy. *Research in Science and Technological Education*, *6*, pp. 167–80.

HAEUSSLER, P. (1987a). Long term effects of science education and their relevance for responsible citizenship. In RIQUARTS, K. (Ed.). *Science and Technology Education and the Quality of Life, Vol. I*. Kiel: IPN; pp. 207–18.

HAEUSSLER, P. (1987b). Measuring students' interest in physics — design and results of a cross sectional study in the Federal Republic of Germany. *International Journal of Science Education*, *9*, pp. 79–92.

HAEUSSLER, P. and HOFFMANN, L. (1990). Wie Physikunterricht fuer Maedchen interessant werden kann. *Naturwissenschaften im Unterricht — Physik*, *34*, pp. 12–17.

HAEUSSLER, P., HOFFMANN, L. and ROST, J. (1986). *Zum Stand physikalischer Bildung Erwachsener — eine Erhebung unter Beruecksichtigung des Zusammenhangs mit dem Bildungsgang*. Kiel: IPN.

HAEUSSLER, P., FREY, K., HOFFMANN, L., ROST, J. and SPADA, H. (1988). *Education in Physics for Today and Tomorrow*. Kiel: IPN.

HIEBERT, E.N. (1962). *Historical Roots of the Principle of Conservation of Energy*. Madison, WI: Department of History, University of Wisconsin.

HINES, J.M., HUNGERFORD, H.R. and TOMERA, A.N. (1987). Analysis and synthesis of research on responsible environmental behavior: a meta-analysis. *The Journal of Environmental Education*, *18*, pp. 1–8.

KESIDOU, S. and DUIT, R. (1991). *Waerme, Energie, Irreversibilitaet — Schuelervorstellungen im herkoemmlichen Unterricht und im Karlsruher Ansatz*. Occasional Paper, IPN, Kiel.

KESIDOU, S. and DUIT, R. (1993). Students' conceptions of basic ideas of the second law of thermodynamics. *Journal of Research in Science Teaching*, *29*, pp. 85–106.

KRUGER, C. (1990). Some primary teachers' ideas about energy. *Physics Education*, *25*, pp. 86–91.

Kuhn, D.J. (1979). Students' attitudes toward energy-related issues. *Science Education, 63*, pp. 609–20.

Kuhn, T.S. (1970). *The Structure of Scientific Revolutions*. Chicago, IL: University of Chicago Press.

Langeheine, R. and Lehmann, J. (1986). *Die Bedeutung der Erziehung fuer das Umweltbewusstsein*. Kiel: IPN.

Lawrenz, F. and Dantchik, A. (1985). Attitudes toward energy among students in grades 4, 7 and high school. *School Science and Mathematics, 85*, pp. 189–202.

Lehrmann, R.L. (1973). Energy is not the ability to do work. *The Physics Teacher, 11*, pp. 15–18.

Lijnse, P. (1990). Energy between the life-world of pupils and the world of physics. *Science Education, 74*, pp. 571–83.

Nussbaum, J. (1983). Classroom conceptual change: The lesson to be learned from the history of science. In Helm, H. and Novak, J.D. (Eds.). *Proceedings of the International Seminar 'Misconceptions in Science and Mathematics'*. Ithaca, NY: Cornell University; pp. 272–81.

Ogborn, J. (1986). Energy and fuel — the meaning of 'the go of things'. In Driver, R. and Millar, R. (Eds.). *Energy Matters*. Leeds: Centre for Studies in Science and Mathematics Education, University of Leeds, pp. 59–66.

Scott, P., Asoko, H. and Driver, R. (1992). Teaching for conceptual change: A review of strategies. In Duit, R., Goldberg, F. and Niedderer, H. (Eds.). *Research in Physics Learning: Theoretical Issues and Empirical Findings*. Kiel: IPN, pp. 310–29.

Shulman, L.S. (1987). Knowledge and teaching: Foundations of the new reform. *Harvard Educational Review, 57*, pp. 1–22.

Solomon, J. (1983). Learning about energy: How pupils think in two domains. *European Journal of Science Education, 5*, pp. 49–59.

Trumper, R. (1990). Being constructive: An alternative approach to the teaching of the energy concept — part one. *International Journal of Science Education, 12*, pp. 343–54.

Trumper, R. (1991). Being constructive: An alternative approach to the teaching of the energy concept — part two. *International Journal of Science Education, 13*, pp. 1–10.

Veiga, M.L., Costa Pereira, D.J. and Maskill, R. (1989). Teachers' language and pupils' ideas in science lessons: Can teachers avoid reinforcing wrong ideas? *International Journal of Science Education, 11*, pp. 465–79.

Watts, M. (1983). Some alternative views of energy. *Physics Education, 18*, pp. 213–7.

Chapter 15

Working from Children's Ideas: Planning and Teaching a Chemistry Topic from a Constructivist Perspective

Philip Scott, Hilary Asoko, Rosalind Driver and Jonathan Emberton

This chapter raises a number of issues relating to planning and teaching science from a constructivist perspective (Driver, in press). The issues are considered through a case study of teaching and learning in the area of chemical change.

The constructivist perspective on learning science asserts that learners interpret and interact with the physical world through their conceptualizations of phenomena. These conceptualizations, which are specific to particular phenomenological domains (for example, force and motion, light) are useful in as far as they 'fit' with phenomena and give some predictive capability; they change and develop through interaction with new physical and social experiences.

From a teaching point of view, the question which arises is how to organize the physical and social experiences in a science classroom so as to encourage development or change in learners' conceptions from their informal ideas to those of accepted school science, a perspective often described in the literature as conceptual change teaching. Other contributors to this volume, including Carr *et al.* and Wittrock, present their perspectives on what is involved in teaching from a constructivist perspective. We argue in this chapter that changes in learners' conceptions come about through processes involving both the personal and social construction of knowledge. The case study described in this chapter illustrates the processes of construction which take place, in a classroom, as a teacher plans a sequence of lessons and then establishes, through appeal to evidence and negotiation with students, the argument for a particular interpretation of phenomena.

This case study is one of a series being undertaken by the Children's Learning in Science Research Group in different topic areas to analyze and develop, through reflection-on-action in classrooms, theoretical and practical guidelines for planning and teaching science for conceptual development from a constructivist perspective. The intention is to analyze the learning demands of particular science topics, to select teaching activities and strategies which are appropriate for promoting conceptual development in those areas and to prepare interpretive accounts of the resulting teaching and learning.

We argue that decisions relating to planning and teaching for conceptual development require the teacher to consider first the nature and status of students' existing ideas and understandings, secondly, the nature of the science learning goals set for the topic and hence, thirdly, the nature of the intellectual 'demand' for the student in developing the science view from existing understandings. These demands vary with different science topics and may involve, for example, differentiation of concepts, the application of conceptual schemes to new contexts, the combining of conceptual schemes into an integrated scheme, the development of a new scheme (Driver, 1990). From such analysis, instructional strategies, appropriate to intellectual demands, can be selected. Instructional strategies reported in the literature have been reviewed elsewhere (Scott, Asoko and Driver, 1991). In this volume White argues for a theory of content. The type of analysis we propose here is not of demands of content *per se* but of the nature of the change required of learners in moving from their prior ideas to those demanded by 'the content'.

The intention of this study is not to provide a detailed account of existing good practice, nor does it claim to offer a 'best way' of teaching this topic. What is laid out here is an analytical account of each of the phases of a theory-into-practice process. The action-research nature of the work acknowledges that we cannot anticipate, through 'a priori' planning, how learners will respond to reaching and offers the means for reflection on, and exemplification of, planning and teaching in action.

Background

This particular case study, which documents the planning and teaching of a sequence of lessons on the topic of 'rusting' with a mixed-ability class of 12–13-year-old students, was carried out in a coeducational 11–18 comprehensive school located in a semi-rural setting in the North of England, in collaboration with a teacher-member of the research group who is also one of the authors (JE). Both the teacher and two of the researchers (PS and HA) were involved in all planning meetings, with the teacher having the final decision relating to choice of particular teaching strategies and activities. The teacher had taught this class in the previous year and enjoyed working with them: 'They're willing to express points of view and are used to different approaches'. The class had two seventy-minute science lessons each week, both of which were taught in the same laboratory. The work on rusting extended over three lessons on successive weeks. Prior to the teaching, two planning meetings were held.

Data Collection

The case provides insights not only into the development of students' thinking during the lessons but also into the choices, judgments and decisions faced by the teacher and researchers during both planning and teaching. All planning discussions between researchers and teacher prior to, and during, teaching were audiotaped. The teacher was also audiotaped throughout the lessons, copies of worksheets were collected, and a record of all blackboard work was made. Two small groups (two boys, Marcus and Robert and four girls, Rachel, Tara, Julie and Emma) were followed in detail through the teaching sequence. All six students were interviewed, individually, prior to the first lesson and again prior to the second lesson. These interviews were audiotaped. Both groups were also audiotaped during the lessons and their activities were monitored by a researcher taking field notes. In addition, all written work was collected from the whole class, including responses to an end-of-topic test question on rusting.

The case

The case is presented in two phases: planning and teaching.

1 Planning Phase

In the case study school, the topic 'rusting' is part of a section on 'substances' and relates to work on chemical change. Rusting was chosen as the topic for this study because it offered an interesting teaching challenge in that students would almost certainly be familiar with the phenomenon but not with its interpretation in terms of chemical change. Furthermore it was considered that the current treatment of the topic tended to underestimate the level of conceptual difficulty involved.

Initial planning focussed upon the three factors referred to in the introduction: the nature of the students' existing ideas about rusting; the nature of the science learning goals set for the topic on rusting and the nature of the intellectual demand involved in developing the science view from existing understandings.

The literature on children's ideas about rusting (DES, 1984; LISP, 1982) suggests that students consider that rusting is caused by water and is worse with salty water. Students are reported to have a range of ideas about the nature of rust including: rust as a protective coating, it stops further decay; rust is the metal breaking up; rust is found underneath the surface of the metal; rust is an impurity within the metal which is brought to the surface; rust is a growing mould or fungus.

The work on rusting was scheduled near the start of the teaching on chemical change and the teacher was keen to use the context to help establish the general 'rule' that new substances are formed when chemical changes occur. The study of rusting was intended to contribute to a developing understanding of chemical change and, with this in mind, two learning goals were agreed:

(i) to appreciate that the factors necessary for rusting to occur are air (oxygen) and iron and water;
(ii) to appreciate that rust is a 'new substance' formed at the surface of iron.

We made the following inferences about likely intellectual demands by comparing existing student ideas and the science learning goals.

The idea that water is necessary for rusting to occur appeared simply to reaffirm existing ideas. However, the idea that air is necessary for rusting was likely to be a new idea for most students. It was a notion likely to be missing from existing ideas, but not necessarily contrary to them. The idea that rust is a new substance formed at the surface of iron was also likely to be a new idea but in this case possibly contrary to existing ideas. Lesson planning was carried out taking into account this analysis of 'students' existing ideas', 'science learning goals' and 'intellectual demands'. Various possible teaching routes could have been taken, all consistent with a constructivist perspective.

The question of whether or not to start teaching with an activity designed to elicit students' ideas about rusting was first considered. It could be argued that we already had information on students' reasoning about rusting from the literature and it was therefore not necessary to elicit this from the class itself. However, it was considered important that the teaching should make links with students' out-of-school experiences of the phenomenon and, after considering a number of ways of doing this, an activity was devised which involved each student taking a nail home and placing it in a location where they thought it would go rusty. The activity was seen to have a number of possible functions:

- to involve each individual in some minimum amount of thinking (in deciding where to put their nail);
- to make each student aware of their own ideas about rusting;
- to relate school science to out-of-school experience;
- to allow the teacher and researchers to find out how these students think about rusting;
- to provide a shared experience within the group;
- to motivate, to get pupils 'on the inside of a problem', to create interest and maybe competition (whose nail is most rusty?);
- to set up an information base, a common reference point for subsequent teaching and learning.

The first four of these relate to the issue of making explicit these students' existing ideas. The next two are concerned with the pedagogical issue of motivation, and the last one draws attention to the information about rusting made available through the activity.

It was not anticipated that any new insights on students' ideas about rusting would emerge from this activity but it was considered that the motivational aspects would be important and useful. The following instructions to students were prepared:

Rusty Nail

Take one of the nails and put it in a place where you think it will go very rusty.

To be completed for homework:

1 Where did you put your nail?
2 What is it about that place which makes you put it there?
3 Why do you think that will make the nail go rusty?
4 What do you think RUST is?

The teacher decided that students would each take a nail home three weeks prior to the first lesson. The nails would then be returned and each mounted on a sheet of paper with the answers to the four homework questions. A display would then be set up with the nails placed in sequence, from least to most rusty, along the laboratory wall.

At this stage, the teacher made the following comment about the 'nails activity':

> 'In a sense you've got their ideas in a box and unless you can lift the lid in some way there's no way for anything coming in or going out and this is a way of lifting it . . .
>
> It's a good way of laying kids open to change. They're absolutely certain that water's going to do it but there's still three weeks of uncertainty that it might not. That is when the insecurity is really important for us in changing things conceptually. If everything is secure and well wrapped up there's no way in. Perhaps that's what you've got to look for in a good elicitation exercise . . . setting that atmosphere of ideas being up for test.'

It was agreed that identification of the factors necessary for rusting should be dealt with before considering the nature of rust and that the 'nails display' should act as a reference point for this. Three possible routes for establishing the factors were considered:

Route 1: Present the 'accepted science view' and relate it to the nails display.
In this, the science view would be put 'up-front'. Students would be introduced to the idea that iron, water and air are needed for rusting to occur. They would then be asked to see whether this made sense in terms of what they found with the nails, for example, why didn't the buried one rust? They would thus need to construct the science view for themselves and relate it to their experiences.

Route 2: Build the accepted science view and relate it to the nails display.
In this, a series of controlled test-tube experiments would be used, with iron samples placed in various conditions (for example, in aerated water, in dehydrated air, etc.), to establish the essential factors. Students would be then be asked to consider whether the results from the test-tube experiments made sense when compared with the nails in the display.

Route 3: Take information from the nails display and work from this towards the accepted science view.
This would involve 'building the science view' from the students' activities and results. Students would be asked to consider whether, based on what they had found with the nails, they could work out what is essential for rusting to occur by identifying and testing possible factors.

Each of these routes attempts to make links between students' thinking and the science view and might therefore be considered to be equally valid from a constructivist perspective. However, it should be noted that in the first two routes, the science viewpoint is introduced to *interpret* student generated data whereas in the third route, the science view is *developed from* student generated data.

The teacher felt strongly that students should be allowed to explore their own ideas, that this would keep them engaged intellectually and so opted for Route 3:

> I don't want to come and suddenly squash their ideas (by presenting the science view) because most kids will think that's what always happens in school, you know, that's what always happens in science teaching. So I've got to try and think of a way round to draw out that other factor (the air) which is seen as a positive thing building on their prior ideas.

It was recognized that the Route 3 approach would be practicable in that a range of possible factors for rusting would be suggested from the nails display and these could then be tested through controlled experiments. The teacher was also aware of the possible need to introduce factors such as air which the students might not consider.

Attention now turned to the nature of rust and rusting and two possible strategies were considered:

Route 1: Present the accepted science theory and relate it to factors necessary for rusting.

Students would be introduced to the idea that a chemical change takes place during rusting: a reaction takes place at the surface of the iron, involving iron, air and water and a new substance is formed.

Route 2: Review the students' ideas presented in the nails display about the nature of rust and work from these towards the accepted science view.

Having developed the notion that iron, air and water are essential for rusting, the students would now re-examine their original ideas about what rust actually is.

In talking through this choice of approaches, the teacher made the following points in support of Route 2:

> Go back to the nails display 'cos they've already sort of conjectured about what rust might be. . . . some of these things you could prepare . . . I mean rust under the surface, you can cut some cross sections, make sure you've got one (a nail) sawn in half . . . I suppose it's sort of crude debunking kind of thing for each of the ideas . . .

In other words, the teacher favoured the strategy of returning to the student ideas presented on the nails display and running simple tests to check their validity. He was already anticipating how the alternative ideas about the nature of rust could be addressed empirically. However, a fundamental concern about this approach related to how the science view would be introduced. The theory that rusting is a chemical reaction between iron, oxygen and water, resulting in the formation of a new substance, is not one that students are likely to generate for themselves. In response to this issue, the teacher argued:

In some ways you've got to have a slot for that (the science view), a need for that. You don't need the right answer unless your idea doesn't work.

The following outline for the rusting lessons was finally agreed:

Lesson 1: Review the 'nails activity' identifying possible factors for rusting. Students devise and set up controlled experiments to test the factors.

Lesson 2: Review outcomes from the controlled experiments and identify essential factors. Return to the 'nails display' and consider students' ideas about the nature of rust. Consider which of the ideas are plausible in view of the findings about factors and how those ideas might be tested.

Lesson 3: Test ideas about the nature of rust with a view to establishing that rust is a new substance formed, by chemical reaction, on the surface of the iron. Consider application of the science view to everyday situations, in the prevention of rusting.

How does this planning relate to our analysis of intellectual demands?

The first learning goal was to identify the factors essential for rusting. Here we could anticipate that water was very likely to be identified as a factor, whilst air or oxygen was less likely to be suggested. The teacher would thus need to anticipate ways in which the role of air or oxygen could be made apparent. There were also likely to be inessential factors identified by students which would need to be investigated and rejected. The investigative approach adopted was designed to enable these types of moves to be made.

The second learning goal was to establish that rust is a 'new substance' formed on the surface of iron. Since these lessons were seen as an introduction to the idea of chemical change, the teaching goal was limited to establishing that a new substance is formed — the process and nature of the chemical reaction were not to be addressed. The idea of a 'new substance' is not very precise, but is likely to require students to reject ideas about rust as coming from inside the iron or being produced by some other surface change involving only the iron. Establishing the idea of a 'new substance' required not only the rejection of such ideas, but also the construction of a new idea. The teacher anticipated the types of activities which could be undertaken to investigate the notions about the nature of rust which students suggested with a view to making them appear less plausible. He also anticipated the activity involving weighing the nails to support the idea that rust is a new substance.

Overall, in this planning, the teacher had to anticipate possible ideas which would emerge and plan possible routes which would establish reasoned arguments to enable students to move in their thinking from their various starting points to the intended learning goals. Planning in this way is demanding. There is no single route to be taken through the lessons and some of the activities have to be contingent on what actually happens in the classroom. The teacher was committed to keeping the students on the inside of the problem so the teaching sequence which was devised had to satisfy two conditions: it had to 'keep faith' with students' reasoning and yet, at the same time, build an argument based on a combination of evidence and reasoning which would lead students to the intended

Table 15.1: Factors identified by students as necessary for rusting to occur. (note: 22 students were present for both the initial nails activity and the final test).

Factors necessary	**Frequence of occurrence (n = 22)**	
	Before teaching	*After teaching*
Air or oxygen	5	20
Water or rain	19	21
Cold	6	1
Salt	1	0
Acids	1	0
Dark	1	0
Light	1	0
Changing conditions e.g. from damp to dry	2	0
Contagion from other rusting nails	1	0
Weathering	1	2
Soil	1	1
Inexplicit	1	0

learning goals. In addition, the teacher had to see that the activities were practicable with this particular class and with the time and equipment available to him.

There are important points to make here about planning lessons from a constructivist point of view. First, a careful analysis of possible routes from student starting points to the intended learning goals is necessary. Activities to support progress along these routes need to be anticipated and thought through ahead of time. However, the actual activities which will be undertaken must be responsive to the development of the reasoning of particular students. Planning has therefore to be thorough whilst acknowledging that some parts of it may prove to be redundant.

2 *Teaching Phase*

The students brought their nails to school on the day before their first lesson and helped to assemble the wall display. The students' ideas, taken from the display, about the factors necessary for rust to occur are given in table 15.1 and their ideas about the nature of rust are given in table 15.2. In general, the ideas proposed were in keeping with those reported in the literature. Before teaching, nearly all the students identified water or rain as a factor affecting rusting. Air or oxygen was mentioned by less than one quarter of the students. The most common suggestions about what rust is, included 'mixing of air and water', 'reaction of cold and wet', 'something covering the surface'.. A small number of students suggested the rust was under the metal coating and others used an analogy suggesting rust was a mould, or 'metal rot' or metal 'wearing out'.

(a) The nails activity

The students had put their nails in various places including: outside (on washing line, on wall, on window-sill); in salty water; in a pool of salt water at night and

Table 15.2: Students' suggestions about nature of rust

Suggestions about nature of rust	**Frequence of occurrence (n = 22)**	
	Before teaching	*After teaching*
Reaction of iron with air and water	1	6
Reaction of air and water	0	9
Reaction with cold and wet	4	0
Mixing of air and water	6	0
Something covering the surface	4	0
Rust is underneath the metal coating and comes out	2	1
Mould	1	0
'Metal Rot' wearing out	2	1
Metal goes brown	2	0
Inexplicit	0	5

out in air during the day; dug into soil; dug into soil for one week then next to rusty nails in tool box; half in and half out of water. The researchers interviewed the target students individually before the first lesson and asked them about their nails.

Marcus had put his nail in a brick outhouse.

R: Why did you decide to put it there?
M: 'Cos I thought it would make it rusty because with damp and water on the floor and coldness. It were on't floor like that and when I took it up it stuck a bit to floor and I took it up and there were rust.
R: So what do you think rust actually is?
M: Like water what gets on the nail and then all condensation, like, digs in and mixes . . . er . . . burns away a little steel.
R: How did you know to put it in a damp cold place?
M: I just thought of it, I just tried to think of best place to put it, to make it rust, like if you put owt metal like a spanner without putting it back in the tool box, that rusts, so I thought I'd put it in the same place as the spanner or something.

These comments provide interesting insights into the status of Marcus's ideas about rusting. He has no problem in identifying the role of water in rusting and uses a form of 'prototypical reasoning' in selecting his rusting location (spanners rust here, put the nail in the same place). It is interesting to note that the condition of 'coldness' is part of the prototype situation and Marcus refers to it along with the need for 'damp and water.'

Robert placed his nail in 2mm (sic) of salty water, in a rusty tin under a leaking pipe on the outside of his house. In his written account he says, 'I put it there because I thought it would make my nail go rusty faster and it would be weathered and wet. I think rust is a reaction when metals and air and water meet'.

R: What made you think about where to put the nail?
Rob: I just . . . I asked me parents about what they thought as well and kind of put them all together.

R: Could you say, well I thought of this bit and me mum and dad thought of that bit?
Rob: Yeah. I thought of the salty water.
R: That was your idea. Now, what led you to think that it might be something to do with salty water?
Rob: 'Cos of piers in the sea. The pier stands on like metal and they rust quick.
R: Well that's sensible enough. And what ideas did your parents give you?
Rob: Well they said about outside and 'em I can't think, that's all.
R: So is the water from this pipe dropping into the tin?
Rob: Yeah, kind of.
R: Why do you think leaving it outside and having the water dropping would cause rusting?
Rob: I think the water mixing with the air . . . I dunno.

One of Robert's parents is a science teacher. Robert associates water and salt with rusting and, through talking to his parents, is vaguely aware that air has some part to play. Robert's nail actually turned out to be the rustiest in the class. Although a naturally reserved boy, Robert was obviously delighted with his rusty nail.

It is interesting to compare the different ways in which Marcus and Robert selected locations for their nails. Marcus placed his nail 'in a place where things go rusty'. He did not readily think of the problem in an analytical way, but rather reasoned by analogy from other prototypical situations. Robert, on the other hand, was already aware of possible factors needed for rusting and set about 'constructing' his location through reference to those factors.

Three of the group of four girls had placed their nails outside. Tara put her nail on the window ledge outside her bedroom because her father had previously told her to keep her bicycle inside to prevent it from rusting and because she thought that rust was a mixture of water and air. Julie put her nail in the roof guttering where she knew there would be water and sunlight and Rachel tied her nail, with string, to a washing line because she thought that rain and air would produce rusting. All three said that the rust on their nails was a coating on the surface only. Emma had kept her nail in a bowl of water on the kitchen window ledge; she thought that water and air together 'made a reaction'.

(b) Lesson 1

At the start of this first lesson the teacher drew attention to the nails display and asked the students with the rustiest (Robert), medium rusty (Tara) and least rusty (Nathan) nails to tell the others about where they had put them. Tara had left her nail on a window ledge and Nathan's nail which had been submerged in vinegar was, in fact, bright, shiny and completely free of rust.

The teacher then made a table on the blackboard with three columns labelled 'Essential', 'Helps', 'Not necessary' for rusting to occur. The students were invited to examine the nails display, suggest possible factors for rusting and then locate each in one column of the table. Responses from the whole class were collated on the blackboard (a copy of the completed table is shown in table 15.3). From this it can be seen that all twenty-six students in the class thought that

Table 15.3: A record of the blackboard writing on 14 March

	Essential	**Helps**	**Not Necessary**
Water	26	—	—
Part in water) Part in air)	20	2	2
Salt	6	19	1
Rusty containers	2	18	7
Outside (weathering)	13	8	1
Vinegar	0	2	24
Soil	0	17	7
Heat	5	18	4
Wet then dry	8	11	7

water was 'essential' for rusting. The notion that air had a part to play was apparent in 'part in water, part in air'. A further popular factor was 'weathering'. Other ideas such as the presence of vinegar or soil were seen to help rusting but were not seen as essential by students. The factors identified on the table were discussed and groups of students were then asked to design controlled experiments to verify or disprove individual factors.

Robert and Marcus worked on the effect of salt on rusting; a factor whose role Robert had commented on in earlier discussions. Three small iron nails were placed in boiling tubes containing salty water, tap water and salt crystals respectively. Emma, Julie, Rachel and Tara opted to test the idea that 'the weather' was needed for rusting. To test this idea they initially decided to put a nail in an exposed position outside and to use a second nail, in an open test tube outside, as a control presumably to reduce the 'weathering'. Thinking they could use only two nails, Tara raised the question of the need for a further control nail inside:

Ta: How're we going to compare all the things, cos there won't be any inside if we have 'with a test tube' and 'without a test tube'.

E: Yes, but we could compare it with everyone else's won't we? (referring to the nails display on the classroom wall).

Ta: Yes, but no-one's going to do an inside one in a test tube and one not in a test tube.

Ra: We'll have to do it.

Ta: Well, we can just tell because, like if it hasn't gone rusty then we'll know that weathering isn't needed, like rain isn't needed.

E: Yes, but what if it's gone slightly rusty?

Ta: Then we'll know that it . . .

Ra: We can tell . . .

E: We can compare it with the nails on the wall (on the 'nails display') because some were kept inside weren't they?

Ta: Yes, good idea.

They then decided that using the nails on the display would not provide a suitable control since those nails had rusted for three weeks, whereas their experiment was to last six days. Finally it was decided that one nail would be left inside and one put on a path outside. In designing their investigation the girls treated 'the weather' which they saw as involving principally rain and sun, as a single variable. As their discussion shows, they were, however, very concerned about the design of their experiment. They considered various possibilities, rejecting them as falling short in some way, and finally agreed that an appropriate control would be a similar nail inside with the experimental one left outside in the 'weather'. They showed an understanding of the function of controls and discussed the significance of different possible outcomes to the experiments. The fact that three of them had put their original nails outside (where they all rusted) was not referred to.

Other groups set up experiments to test for the effects of half air half water, acid, heat, soil, rusty container. All of these experiments were left until the next lesson in the following week.

After examining the range of experiments organized by the students, the teacher decided to set up two further experiments to ensure that all necessary information, for isolating the factors essential for rusting, would be available to the class. This involved placing one nail in oxygen-free water and a nail made of copper in aerated water.

(c) Lesson 2

Prior to this lesson, the researcher spoke to Robert about his experiment. Robert predicted that the order of rustiness would be salty water, water, salt (which wouldn't rust at all). In fact, the nail in the water looked most rusty, although a lot of rust appeared to have 'fallen off' into the salty water. The researcher referred Robert back to the table on the various factors involved in rusting:

> *R*: As far as you're concerned, have you anything in mind that's essential (for rusting to occur)
> *Rob*: Water and air, half in half out.
> *R*: Have you anything which you think is not essential or not needed?
> *Rob*: Salt is not needed, might help, I don't think it's needed.

Robert was able to state with confidence that, 'salt is not needed' but the issue of why salt did not produce accelerated rusting in the boiling tube, when it appears to at the seaside, remained unresolved.

The girls were interviewed in pairs about the outcomes of their investigation on the effects of weather. The nail which had been left outside was covered in rust and small patches of rust were visible on the nail left in the laboratory. Both pairs of girls interpreted these rusults in terms of 'weather' being important for rusting. None of them had expected that the nail left inside would rust and explained the rust patches by saying that condensation from a nearby window was the likely cause.

At the beginning of the second lesson, groups collected their experiments and inspected the nails. The teacher then give each pupil a prepared grid with the possible factors involved in rusting listed on it. Students first completed the table for their own experiment, identifying prevailing conditions, and then collected similar data from other groups. The teacher also reported the outcomes of his two experiments to the class. The nail in oxygen-free water had not rusted and neither had the copper nails.

When students had completed their tables, the teacher put a copy of the table on the board and led a class discussion to identify the essential factors from the experimental results.

T: I want people to . . . think about which things really seem to be showing up as causing things to go rusty and which things are perhaps red herrings — which don't seem to be causing it . . . (hands go up) Robert?
R: Water.
T: Would you all agree that the water seems to be very, very important?
(*This was clearly in keeping with the results that pupils obtained from their experiments*).
Any more? OK Jonathan.
J: Oxygen.
S's: Yes.
T: The oxygen column at the end there seems to be showing up as very important.
(*This was in keeping with the results obtained from the experiment set up by the teacher which had been noted by all the class*).
We can't actually identify oxygen as being a separate thing, but we all know from our work before that oxygen is definitely part of air isn't it . . . if there's air we assume there must be oxygen all right?

The teacher then turned to other factors on the table which had attracted a large number of ticks: salt, weathering and soil. He first took the case of the effect of salt and asked Robert and Marcus to tell the class what happened in their experiment.

M: The water one went rustier (than the one in salt water).
T: The one which Marcus hasn't mentioned which I know he did which makes this even more convincing is . . .
M: We filled it a little bit with salt and put a nail in.
T: Right. So they've just got salt on its own with no water . . . and there was no rust there. So that must prove that the salt on its own cannot cause the rust to be there, all right . . . you'd support that sort of view, wouldn't you?

The next factor suggested was that of 'weather'.

T: Yes, if it's out in the weather it does go rusty. What is it about the weather which you think is causing it to be rusty though? Richard?
R: Oxygen and water.
T: You've got the air and you've got the water there, well done. And it might well be that if it's out in the weather it just gets more water alright? So that's why that might be the case.

Here the teacher is asking the class to consider the everyday 'weather' factor as being the combined influence of basic variables air and water. He then continues:

T: What about the soil? Why do things go rusty when they're in the soil? Not because of the soil, James?
J: 'Cos there's water and oxygen around there.
T: Water and oxygen around there. This is good isn't it? In other words we're beginning to see that, when you say that the soil caused things to go rusty at home I can understand why you immediately thought, yes, it's the soil that's doing it. But now we're beginning to look a bit deeper we can see that it's the factors involved with that soil. Alright, so each of these things we're beginning to see a bit further into it. So I'm going to make a statement now which I hope that we could all agree with, right, about what I consider to be the three things which are essential if rust is going to, er, take part. We must have water, we must have air, if you like . . . more specifically oxygen and the one which nobody mentioned last week . . .
S: Heat.
S: Iron.
T: Well, the heat we've dismissed really because we've no ticks in there. The iron.
S: You've got to have the nail.
T: Yes, but not just any old nail.
S: An iron nail.
T: It's got to be an iron nail, and that's the key thing alright. So something which nobody has picked up on earlier has come out very, very strongly there.

In this extended class discussion, we see the teacher developing an argument with the class for the factors that are essential for rusting. It is interesting to note he makes the strong claim that, 'It's got to be an iron nail' for rusting to occur. The information available to the class at this time (that iron nails rust, but copper nails do not) can provide supportive, but not conclusive, evidence for this claim. The teacher is moving beyond the limits of available evidence as he helps the class to identify the essential factors. He also reinterprets their complex factors in terms of these essential factors.

Having established the factors essential for rusting, attention turned to the question, 'What is rust?' The teacher had grouped on the blackboard the various ideas suggested earlier by the students about the nature of rust (shown in table 15.2) and now considered which of these would be worth testing:

> What I want to do now is to be a bit drastic, because I want to sort some of these ideas out in terms of what we've decided just now. We've said that water and oxygen and iron have just got to be there, alright? So let's go through these with a piece of chalk and decide on the ones which don't involve water, oxygen and iron. And then I think it'll be fair to say that if they don't involve those things then we have to say those aren't what we want, OK? So, let's go through them one at a time and see.

By using this approach, all but three possibilities for the nature of rust were eliminated; these were that rust might be: the product of a reaction between iron, water and air; present under the surface of the metal; a growing, living mould (which needs water and oxygen to survive). In the time remaining, students were asked to discuss in groups how they might test each of these ideas.

(d) Lesson 3

The lesson began with the teacher asking the students for suggestions about how they might test the three ideas, identified in the previous lesson, about the nature of rust. This led to various activities; nails were sawn in half (to search for rust under the surface) and rust was examined under a microscope (supposedly to inspect for living characteristics). Rust could not be found 'inside the iron' nor did it appear to be living. The teacher also asked the students to consider how they might test the chemical reaction idea. The suggestion was made that such a reaction would result in 'an increase in weight because there's a new substance'. The teacher had anticipated this line of reasoning and had earlier weighed a bundle of nails (with the help of two girls from the class) and then placed them in a gas jar with a wad of wet cotton wool. The nails were now very rusty and the teacher reweighed them in front of the class to find a significant increase in mass (27.78g to 28.29g).

T: So something extra is now sitting on those nails to give it that extra weight.
S: It's rust.
T: We know it's rust definitely (laughter) the point is that a lot of the ideas we've had, said something . . . if there's extra weight there it means that there's something as well as the iron there. Alright, so what two things have we said are likely to be there as well as the iron? Remember from yesterday's work.
S: Water and oxygen.
T: Water and oxygen. So it looks, from the evidence there is, as if the water and oxygen have linked up together with the iron to be sitting there as a new substance, as a new chemical and that's why the weight has gone up because a new chemical has been made. And the water and the oxygen wasn't on the nails originally and it's now there, connected up as rust and it affects the weight. It's not much is it? But it's certainly noticeable.

The teacher then summarized what had been done in the three lessons.

> We've had three lessons where I think we've really made a lot of progress. I really do think we've made a lot of progress because we've definitely decided the three things we think make, cause, rust, OK, the iron and the oxygen and the water. And we've come to a fairly good decision, I think, that there's a new substance that's been made there and when we get that extra weight it's because those three things have reacted to make a new substance . . . And finally, we've just looked at some of the ways you can stop things going rusty and it's dead easy. If you can stop either the iron being there, stop the water being there or stop the oxygen being there or more than one of those things, then there won't be any rust. You've got to have three out of three and unless you've got all three there's no rust.

Figure 15.1: Post test question on rusting

Iron Railings

Calderdale Council decided to repair some old iron railings. When the railings were inspected it was found that: the parts of the railings above the ground were very rusty
the parts of the railings under the ground were quite rusty

1 Explain why there should be this difference in rustiness:

____

____

There really was a lot of rust on the railings.
2 Explain where the rust had come from:

____

____

The rust was cleaned off the railings.
3 What can be done to stop the railings from going rusty again?

____

____

4 Explain how this would stop further rusting:

____

____

After these lessons on rusting the class continued with further work on chemical changes. Six weeks later the students were given a standard school science test which included one question on rusting. This is shown in figure 15.1. Students' responses were analyzed to identify which factors they noted as affecting rusting and what they considered the nature of rust to be. These are entered in the 'after teaching' columns in tables 15.2 and 15.3. From these tables it can be seen that there was considerable change in students' stated ideas about the factors necessary for rusting to occur, with nearly all students after teaching identifying both air and water. Other factors mentioned previously are now hardly mentioned at all. Students' stated ideas about the nature of rust have also changed as a result of teaching though to a lesser extent than the knowledge of the factors.

The teacher later commented that he had been pleased with the way in which the rusting lessons had gone. He felt that the students had been much more engaged with the topic than had previously been his experience with other classes. He also commented that the work on rusting became a very useful reference point, for this class, in subsequent lessons on chemical change.

The teacher felt that the approach to planning the lessons, using a 'schedule' which makes explicit reference to children's thinking, science aims and intellectual demands, had been useful.

How was Constructivist Teaching Implemented in this Case?

The students' ideas elicited through the nails activity were largely as expected. There was one feature of their reasoning, however, which only became apparent during the initial lesson. In many cases students' thinking about where to put their nail was governed, not by analytical model of factors which need to be present to produce rust, but by a form of prototypic reasoning whereby they referred to cases of rusting familiar to themselves. The reasoning of Marcus is a good example of this. He thought of the problem by analogy with what he knew happened to tools left out from his father's tool box. By contrast Robert chose the place for his nail by explicitly identifying factors which he considered important and then constructing an environment which maximized these.

This prototypic reasoning interacted with instruction when students were asked to identify the 'factors'. In this situation 'weathering' and 'soil' were suggested as factors possibly because they were part of students' prototypes of rusting. Here, isolating the factors essential for rusting involves not just the addition of new factors (like air) and the rejection of others, but a change in the type of reasoning from one which is situated and 'contextually bound' to a decontextualized analytical perspective. The way the teacher dealt with this shift, as we saw in the case, was to allow the investigation of complexes of factors such as 'weathering' and then, through argument, to show the students that each was an example of possible combinations of the essential factors. In other words, he offered a 'reinterpretation' of their findings.

The strategy of letting students identify the factors they thought would be influential in rusting and carrying out controlled experiments to test them had a number of problematic features to it. As we have just noted some of the 'factors' identified by students involved undifferentiated complexes which needed further interpretation by the teacher, even when a controlled experiment had been appropriately conducted. In some cases, experiments appeared 'controlled' to the students but, through lack of knowledge, proved not to be so. This produced results which were hard to interpret. A case in point was the experiment on the role of salt carried out by Robert and Marcus. The experiment they set up included an iron nail in a test tube of salty water, a test tube of water and one in dry salt. In this case the role of dissolved oxygen was not considered nor controlled for in the test and the results which showed that the nail in the water was rustier than that in salty water was an unresolved puzzle for Robert.

This experiment illustrates the problem of using an experimental approach to investigate the role of individual factors when the phenomenon in question involves interacting variables, some of which students do no know about (for example, air/oxygen) and therefore do not consider in their design. In this case the students did not identify all the necessary factors and the teacher, noting this, set up illustrative experiments for consideration alongside the student generated ones.

Similar issues arose in exploring the nature of rust where activities suggested by the students focussed on 'alternative' ideas such as rust under the surface or rust as mould. The activity involving the weighing of the nails was purposely included by the teacher to enable him to make a case for a 'new substance'. The process of investigating personal ideas and theories may lead students to reflect upon and question them. At the same time, it is unlikely to lead to the scientific view. As we saw in this case, the teacher needed to augment the students' own investigations to provide evidence for the scientific view.

How then did the scientific view get established in the lessons? The teacher's role here was critical. As we have already indicated he analyzed the investigations that the students undertook and augmented them as necessary to provide essential evidence. However, his function in establishing the science view was not simply one of producing the empirical evidence. The issue of what evidence to focus on and how it should be interpreted was clearly structured by the teacher when the results of the experiments were reviewed. In this way the scientific view was established and made to seem reasonable to students through discussion. The discussion of the results of the 'factors' experiments was a clear illustration of this process. What we see happening here is the construction by the teacher, of an argument in support of the science view. In developing the argument he draws on evidence from both the student's and his own experiments and deals with alternative proposals in a logical and open way. In the case of establishing that iron is essential to rusting, we have seen that the teacher actually goes beyond available evidence to support the developing argument.

The extent to which students are able to follow the argument and construct it for themselves, however, is a distinct though related issue. It was notable that during the class discussion on 'factors' the teacher was drawing on suggestions from the class and in this way obtaining some feedback as to whether they followed the argument. He also used 'checking' questions such as 'do we all agree that . . .?' It is perhaps important to note that, during discussion of the nature of rust activities and weighing the nails, the teacher did not interact with the class to the same extent in establishing the argument that rust was a 'new substance'. We are left with little evidence of the extent to which this argument was followed by the students and from the post test it appears that fewer students did in fact construct this notion.

Teaching Science From a Constructivist Perspective

Consideration of this case leads us to draw attention to four aspects of teaching which is informed by a constructivist view of learning.

1 There is no unique method or instructional route for teaching a particular topic from a constructivist perspective. We would argue, however, that a central focus of planning such instruction should be in comparing the students' and the accepted science point of view, thus providing insights to the intellectual demand, for the learner, of developing the science view. Such analysis may identify common elements in student and science perspectives and thereby suggest an instructional strategy which involves exposing students' initial thinking and using that as a starting point to teaching, for example, by using bridging analogies as suggested by Clement *et al.* (1989). In other cases no such common elements may exist and instruction may then be based upon helping students to construct a science view before returning to reinterpret phenomena previously seen in other ways (Rowell and Dawson, 1985). Whatever the instructional approach adopted, the teaching should aim to support students in making links between their existing conceptions and the science view.

2 Learning science involves not only coming to terms with new conceptual structures but also involves developing a new rationality for knowledge. This rationality values decontextualized rather than situated knowledge; it values explanations which are generalizable to many contexts rather than those which are limited and ad-hoc in nature; it demands internal consistency of theories. Science instruction needs explicitly to acknowledge this new basis of rationality rather than leave it as an implicit feature of science learning. Science learning, viewed from a constructivist perspective, involves epistemological as well as conceptual development (the chapter by Carr *et al.* in this volume (11) addresses these points in detail).

3 The teaching involves establishing an argument for the science view which is likely to involve empirical findings but goes beyond these in helping students to construct the particular 'ways of seeing' adopted by the science community. The teacher develops a narrative of introduction to the science view in which supportive evidence is drawn upon preferentially and new modes of expressions are rehearshed through discourse in the classroom. Learning science involves socialization into a particular way of looking at the world. It is not a matter of discovering 'how the world really is'; the science view is simply not there to be 'seen' in the real world. This highlights a very important distinction between discovery learning and constructivist approaches to learning. Since the science view is itself socially constructed within the science community, learning science requires students to be socialized into a 'new way of seeing'; they need to be enculturated into the science community.

4 Teaching informed by a constructivist perspective recognizes that both practical activities and the discussion of these may be interpreted by students in ways which differ from those intended. Even when arguments have apparently been clearly developed through classroom discourse, this does not mean that individuals have made sense of them. Teaching must involve a process of regular feedback and checking to identify the reasoning students are using so that teaching activities can be adjusted accordingly.

Note

In the transcripts, R refers to researcher and T to teacher. Particular students are identified by their initials or an abbreviated form of their name. In general class discussion students are identified by S.

References

Clement, J., Brown, D. and Zietsman, A. (1989). Not all preconceptions are misconceptions: Finding anchoring conceptions for grounding instruction on students' intuitions. *International Journal of Science Education*, *11*(5), pp. 554–65.

DES. (1984). *Science in Schools. Age 15: Report No 2.* Assessment of Performance Unit. London: HMSO.

Driver, R. (1990). Theory into Practice: A constructivist approach to curriculum development. In Fensham, P. (Ed.). *Development and Dilemmas in Science Education*. London: Falmer Press, pp. 133–49.

DRIVER, R. (in press). Constructivist approaches to science teaching. Paper presented in the seminar series *Constructivism in Education*. Mathematics Education Department, University of Georgia: Lawrence Erlbaum Associates.

LISP. (1982). *Reactions*. Working Paper No. 37. Hamilton, New Zealand: Science Education Unit, University of Waikato.

ROWELL, J.A. and DAWSON, C.J. (1985). Equilibration, conflict and instruction: A new class oriented perspective. *European Journal of Science Education*, *4*(4), pp. 331–44.

SCOTT, P., ASOKO. H. and DRIVER, R. (1991). Teaching for conceptual change: A review of strategies. In DUIT, R., GOLDBERG, F. and NEIDDERER, H. (Eds.). *Research in Physics Learning: Theoretical Issues of Empirical Studies*, Kiel, Germany: University of Kiel, pp. 310–29.

Chapter 16

States of Matter — Pedagogical Sequence and Teaching Strategies Based on Cognitive Research[1]

Ruth Stavy

It is a broadly accepted assumption that human beings construct their knowledge by a continuous process of interaction between the individual's cognitive system and his/her physical and cultural environment. This chapter focuses on the question of how children construct specific knowledge about the concepts, solid and liquid, and how different content-related factors (experience, perception, language, etc. — see White, chapter 18 in this volume), or learner-related factors (age, instruction, etc.) affect this process. This is done with the belief that if we understand how children represent these concepts, how they construct their knowledge about them, and how the above-mentioned factors affect this process we should be able to improve our methods and timing and sequence of teaching them.

The chapter will first refer to the concepts of solid and liquid, and of change of state, and to their respective places in the curriculum. It then discusses our studies on the development and learning of these concepts. (Stavy and Stachel, 1985a and 1985b; Stachel and Stavy, 1986; Stavy, 1987). Finally the information obtained in these studies will be used to: (a) rationalize the timing and sequence of teaching these concepts; (b) suggest particular teaching strategies to deal with certain conceptions manifested by children, especially by supplying suitable examples and non-examples in order to help students extend or limit their knowledge to the appropriate boundaries of each specific concept; and (c) provide points for discussion with students with regard to problematic aspects of these concepts.

Solid and Liquid and their Place in the Curriculum

As many of the concrete substances which surround us in everyday life are not pure it is often difficult to categorize them into two separate groups of solids and liquids. Even some of the pure substances do not always fit this classification. However, the concepts of solid and liquid are widely accepted in science and in science education (see Driver *et al*., 1987; and Fensham, chapter 2 in this volume). They, and the topic of change of state, are usually taught in junior high school as an introduction to the particulate theory of matter. The physical attributes of these concepts serve as part of the phenomena on which the particulate theory is based.

Junior high school textbooks usually assert that matter can exist in three different states: as solid, liquid and gas. Solids are defined as having a definite volume and shape, liquids as having definite volume but no definite shape, and gases as having neither definite volume nor definite shape. It is usually also mentioned that all matter shares some attributes: it has mass (weight) and occupies space. These physical attributes, in turn, the textbooks explain with the particulate theory of matter.

Children's Conceptions of Solid and Liquid

The development of concepts can be studied from two different aspects (Bourne, Dominowski and Loftus, 1979). One is by studying the nature of children's classifications of different items according to the concept under study. The other is by studying the nature of children's explanations or definitions of the concept. Explanation or definition of the concepts solid and liquid, can again be given on two levels, one related to the physical attributes at the macroscopic level and the other to the particulate theory of matter at the microscopic level. These two aspects were used in a study to follow the development of the concepts, solid and liquid, with age and with schooling from informal intuitive concepts to formal scientific ones. Children of different age groups and in different instructional programs, unschooled women (who had never attended school), and schooled adults (students in teachers' college) were asked to: (i) indicate how two solid objects are similar and how two liquid objects are similar and to explain the concepts solid and liquid; and (ii) to classify different objects into solids and liquids.

Solid and Liquid — Definition of Concepts

Liquid

Usually children represent liquids by way of two mental images or models: either by a prototypical exemplary substance — *water*, or by a typical common behavior or property — *pourability*. How are these representations reflected in children's behavior? When children are presented with two different typical liquids — for example, tea and perfume — and are asked if there is anything common to both, practically all children from the first grade (6-years-old) on, including 20 per cent of kindergarten children (5-year-olds) and 50 per cent of unschooled women recognized similarity between them. It seems that different liquids have *salient common perceptual* properties that children at a very early age recognize without direction or guidance. The younger children (up till third grade) explained that liquids were similar because they were *made of water* ('contain water', 'they are both water', 'similar to water'), or because they both *poured*. Use of the word *liquid* to describe similarity between liquids lags behind the ability to recognize it. It increases gradually from 56 per cent in the first grade to 100 per cent in the seventh grade, while only 4 per cent of the kindergarten children used the word *liquid*. This finding differs from others related to the same issue (Dickenson, 1982). The use of the word *liquid* was found to be very poor among English-speaking children (only 14 per cent in first grade compared to 56 per cent of the Israeli children, and 50 per cent of the seventh graders as compared to 100 per cent of their Israeli peers). In Hebrew, the noun for 'liquid' (nozel) is the same as the verb 'pour'. It seems that the *linguistic factor* helps Israeli children describe liquids. In the younger age groups, when the word *liquid* is probably not yet available, Israeli children use other descriptions. These descriptions relate either to a typical exemplar of liquid (based on visual similarity), or to a typical property characteristic of liquids (based on concrete actions with liquids). Once the word *liquid* is available to children, they tend to use it. However, as we will see later, the idea that all liquids are water, which prevails in grades 1–3, lingers in older children as well.

When children were asked to explain what a liquid is, most of them in all age groups could give a relevant definition of liquid. As might be expected, the majority of them said liquids could be poured. A second property given by all age groups though less often, was the similarity of liquids to water. Children may simultaneously hold the two different models for liquid (pourable, water) and use them in different situations. About 40 per cent of seventh grade students, *after studying* 'The Structure of Matter' (see page 229 for description) defined liquid by means of school-learned physical properties such as flowing and changing shape, and only about 25 per cent of these seventh graders related it to the theoretical microscopic level.

What are the consequences of representing liquid as: (a) something that pours? (b) something resembling water?

(a) Representing liquid as something that pours may cause overgeneralization, i.e., including other pourable substances that are not liquids, such as powders.
(b) Representing liquid as something watery may sometimes cause undergeneralization, i.e., relating to all liquids as if they were water and so excluding heavy oils, treacle, etc. Another outcome of representing liquid as water will be described later with respect to the candle-melting problem.

Solid

Usually children tend to represent solids as something hard. When presented with two different typical solids — for example, a piece of rock and a wooden stick — and asked if they have anything in common, it seems that observing similarity between solids is not as easy as observing similarity between liquids. Most children until the third grade are not aware of the similarity between solids. From grade 4 on there is a gradual increase in this awareness. It seems that the common properties of different solids are much more abstract and difficult to grasp than those of liquids, and are not generally recognized until quite late. Use of the word solid to describe similarity between solids lags behind the ability to recognize it and is much less frequent than the use of the word liquid. Only about 30 per cent of grade 5–7 children used the word *solid* to describe similarity between solids. The gap between observing similarity and using the word *solid* is much larger than the parallel gap relating to liquids. Instead of using *solid* to describe similarity between solids, children used other descriptions, for example, *hard, inanimate, strong, unbreakable.* In contrast to liquids, there were no descriptions relating to a prototypical exemplar of solids, but all descriptions did relate to typical properties of solids. The most common property used by the majority of children to define a solid was 'hard'. The second property mentioned by most age groups was 'not liquid'. Seventh grade students, *after studying* 'The Structure of Matter', used 'hard' (44 per cent), school-learned physical properties such as 'constant shape', 'has weight' and/or 'volume' or 'crystalline' to define a solid. Only 25 per cent of these seventh graders referred to the theoretical microscopic level in their definitions.

What may be the consequences of representing solid as something hard? Sometimes it may lead to undergeneralization, i.e., including only a sub-group of solids — rigid matter — and leaving out sub-groups such as non-rigid solids (elastic and plastic solids), and powders.

Solids and Liquids — Classification of Substances

So far we have seen that the concepts solid and liquid — the only concrete-tangible forms of matter — develop independently at different ages. Solid is conceived as something rigid and resistant to change, and liquid as something that can be poured or as something similar to water. A question arises as to how children will classify actual solids and liquids. Their definitions necessarily lead them to classify these substances into more than two groups. There will also be substances which according to their criteria belong to neither group, for example, aluminium foil, which is neither hard nor pourable; or honey which does not pour fast and is not hard. On the other hand, there will be substances that might be related to both groups, for example, sand which pours and is not watery. Indeed, our findings indicate that children did classify substances in more than two groups. They relegated all the substances which they did not classify as either solids or liquids to an intermediate group designated — 'both liquid and solid', or 'neither liquid nor solid'.

A more detailed description of children's classification activities is presented below.

Kindergarten children and unschooled women

Unschooled women and kindergarten children demonstrated a lack of success in classifying objects as solids or liquids.[2] Kindergarteners put liquids into the liquid group in 65 per cent of the cases and solids in the solid group in 33 per cent of the cases; unschooled women related liquids to the liquid group in 57 per cent of the cases and solids to the solid group in 22 per cent of the cases. Both groups used categorization systems based on association, mainly according to use or immediate visual similarity and they were not consistent in applying their criteria. Unschooled women as well as kindergarteners have a lot of experience with plenty of solids and liquids, but this experience is apparently insufficient for the development of the concepts of solid and liquid. It seems (as will be shown later) that children who are going through the educational system develop the core concepts of solid and liquid as a consequence of acquiring general skills of observation and classification in the elementary school years (Stachel and Stavy, 1986).

School children

From first grade on we observe the use of a consistent and well-organized classification system. This is accompanied by a considerable increase in the percentage of children who correctly classify liquids (about 95 per cent), and a parallel increase in the ability to classify solids which reaches a success level of about 60 per cent. This latter level remains unchanged until the seventh grade, when it starts rising until it reaches 80 per cent in the ninth grade. The difference between the ability to classify liquids and the ability to classify solids matches the previously described findings concerning the ability to identify similarity between solids and doing the same for liquids. In both cases children demonstrated a better understanding of the concept of liquid than of solid.

The rigid solids were correctly classified by the majority of children from first grade up. Non-rigid solids were less successfully classified by around 50 per cent of all age groups, beginning with the first grade, with no major change observed with age until seventh grade, and then reaching 100 per cent in ninth grade (after studying 'The Structure of Matter'). The materials in this group (elastic or plastic objects) easily change shape, causing a considerable number of children not to relate them to the solids, but to an intermediate group. Most children from all age groups were unsuccessful in classifying powders as solids. Powders were mostly classified as an intermediate group or as liquids. These results indicate that practically all school-children have some knowledge of the concept of solid, but this is much more limited than the concept held by a scientist and includes mainly rigid solids, excluding non-rigid solids and powders. For many children, *solid* thus refers to a rigid material with an unchangeable shape or state.

As stated above, powders were problematic for children of all ages. Powders are materials which take on the shape of their container, and generally behave as liquids (unless regarded as separate grains). Their pourability adds to children's tendency to include them under liquids. It seems that the finer its grains, the less likely the powder is to be classified as a solid. Classification of powders with liquids decreases with age and is almost absent in the sixth and seventh grades. Powders were classified as intermediate groups in about 50 per cent of the cases until grade 7 and then dropped to 20 per cent of cases in the ninth grade.

The justifications for the classification of powders as solids derive from two sources: first, from grasping the fact that powder is 'not liquid'; second, from the understanding that powder consists of small, solid grains.

The Effect of Teaching on Children's Conceptions of Solid and Liquid

The data concerning classification of solids and liquids presented above indicate developmental changes across a wide age range with distinctive periods of progress: the first — from kindergarten to first grade when we witness a dramatic change in the ability to classify liquids and rigid solids, and the second — from seventh to ninth grade (seventh grade students were tested before they studied topics related to the states of matter), when a considerable increase in the ability to classify non-rigid solids and powders occurs. It can be assumed that this picture reflects the effects of age and school learning on children's conceptions of solid and liquid. Many studies now, however, have indicated that the ability to classify according to formal criteria is affected mainly by school learning and to a much smaller extent by age. The results obtained with unschooled women also suggested that these subjects could not classify materials into solid/liquid, and that, instead, they tended to use functional criteria.

In the light of these findings we asked whether teaching might affect the development of the concepts of solid and liquid during those periods in which we witnessed remarkable change.

Kindergarten Children

In order to find out what in the kindergarten program is likely to contribute to the development of the solid/liquid concepts, the effects of two teaching programs on this development were compared.

One of the programs is the kindergarten program commonly used in Israel. The second program included an additional study unit — the MATAL Early Childhood Program (Stachel, 1986). This additional unit deals with three central issues relevant to the present study: (a) Properties of objects and materials such as shape, color, smell, taste, texture, etc.; (b) differentiation and discrimination between these perceptual properties; and (c) classification skills.

It is important to note that neither program dealt with the concepts solid and liquid, or with the states of matter. It was therefore interesting to investigate whether teaching centered around acquaintance with properties of objects and materials through both sensory activation and classification activities would enable children to observe properties of materials in an area not included in the program; and moreover whether such teaching would improve the understanding of the concepts *solid* and *liquid*, as compared with children's achievements in the regular program.

Children's Definitions of Solid and Liquid

Liquid

During the kindergarten year a significant development took place in both groups of children — those who studied according to the MATAL program, and those who studied according to the regular program — in their ability to recognize similarity between two liquids presented to them (tea and perfume). It seems that the MATAL program significantly accelerated the development of this ability (MATAL children's ability to recognize similarity between two liquids increased from 17 per cent at the beginning of the school year to 83 per cent at the end of that year, while the regular program children's ability increased from 8 per cent to 58 per cent). However, there was no difference between the two groups in the nature of their descriptions of the similarity between the two liquids. There was no difference in the use of the word *liquid* between the groups at the beginning of the year (only 4 per cent of the children used the word liquid), while at the end of the year 62 per cent of the MATAL program children used the word liquid as compared with 37 per cent of the regular program children. These findings indicate the accelerating effect of the MATAL program on these abilities, but the use of the word *liquid* to describe similarity between liquids still lags behind the ability to recognize it. The children's ability to define the concept *liquid* did not change in the course of the year.

Solid

Recognition of similarity between solids is more difficult than between liquids, and even the MATAL program did not have a significant influence on the children's achievements. Here too, use of the word solid to describe similarity between solids, lags behind the recognition of similarity, and is used much less than the word liquid (only 12 per cent in the MATAL group). With regard to relevant definitions of solid, we observe a significant advantage on the part of the MATAL group (there was no significant difference between the two groups at the beginning of the year). This indicates that the program significantly enhanced the ability to explain the concept solid, even though it is difficult to point at a substantial change in the nature of the explanation.

Classification of Solids and Liquids

Liquids

Thin liquids were correctly classified by the majority of the children in the two groups. The percentage of children who classified thin liquids with solids is very small even in the first test (at the beginning of the year — prior to the learning), and remained the same in the second test (at the end of the year — after the learning) without any difference between the programs. Only around 15 per cent of the children related thin liquids to the intermediate group (neither 'solid' nor 'liquid').

A reasonable percentage of the children in the first test (around 60 per cent) related dense liquids to the group of liquids. This finding also supports the existence of an intuitive perception of the concept 'liquid'. With the input of the MATAL program, a *decrease* is observed in the number of children who classified dense liquids with liquids and a significant difference is found between the programs in the second test. Parallel to these findings, there is an *increase* in the number of MATAL children who related dense liquids to the intermediate group. However, no similar changes occurred with regard to the percentage of relating dense liquids to solids.

We can assume that this decrease and this increase stem from learning focused on observation which yields a finer distinction between the two groups of liquids, on the one hand, and from the idea that liquid is only a fast pourable material similar to water, on the other. These changes in relation to liquids point to a transition from a diffuse, ill-defined knowledge of the concept, 'liquid', to a more defined approach, albeit still limited in scope.

Solids

With regard to the percentage of children who succeeded in classifying rigid solids, there was no difference between groups in the first and second tests (both around 40 per cent). Around 35 per cent of the children related rigid solids to the intermediate group. There was no change during the year. Regarding the relation of rigid solids to liquids, around 20 per cent related rigid solids to liquids in the first test. In the second test a significant difference was observed between the programs with a decrease in the percentage of MATAL children who classified rigid solids with liquids. This finding also supports the assumption that the MATAL program helped improve the perception of the concept, 'liquid', and this was expressed by not including these materials.

Around 30 per cent of the entire population related non-rigid solids to the group of solids with no difference occurring between the groups during the year. In the first test 55 per cent of the population related non-rigid solids to the intermediate group. A significant difference between the children occurred at the end of the year when more children from the MATAL program related non-rigid solids to the intermediate group and fewer related non-rigid solids to liquids. This result supports the view that the concept of liquid also crystallizes by means of the rejection of non-rigid solids. At the same time, no parallel extension of the concept 'solid' is observed as children do not classify non-rigid solids with the group of solids.

Success in classifying powders was extremely low. Only around 20 per cent of the children put powders with solids in the first test, and there was neither a significant difference between the groups at the beginning of the year nor at its end. In contrast, a significant difference between the groups did appear in the second test when more MATAL children related powders to the intermediate group and, in parallel, fewer related powders to liquids. This finding also supports the earlier ones indicating the impact of learning on the development of the concept 'liquid'. A similar achievement to that of the MATAL group after learning appeared in the previously mentioned study with school children aged 10. This fact attests to the enhancing influence of teaching on the development of the concept. No similarly clear developmental trend can be observed with regard to solids, though from the children's justifications of their classification the beginning of a relevant relation to the concept can be recognized.

In the first test the majority of children in both groups did not possess a system of classification related to the concepts solid and liquid. Classification was generally associative. At the end of the year MATAL children showed a tendency to use a classification system indicating some development both regarding the concepts and the ability to classify: they began using more adequate criteria for the concepts, 'solid' and 'liquid', and performed the classification more consistently and systematically as compared with both the initial state and the achievement of the control group. These findings point to the importance of fostering skills of observation and classification as factors which may contribute to the enhancement and advancement of development of these concepts.

Seventh Grade Students

As mentioned above a dramatic increase occurs in children's ability to classify non-rigid solids and powders, during the shift from seventh to ninth grade. The seventh grade curriculum in Israel contains instruction in the concepts solid, liquid and gas under the topic, *The Structure of Matter* (Orpaz and Ben-Zvi). Here the children are taught the characteristic physical properties of solids and liquids. These concepts are taught formally, and instruction is neither based on nor does it refer to, children's prior intuitive knowledge or the specific difficulties they have in understanding them. After being taught the properties of solids, liquids and gases (usually in the reverse order of presentation, see Fensham, chapter 2 in this volume) the students receive a theoretical explanation of the states of matter based on the particulate theory. In a revised experimental program, *Inside Matter* (Orpaz, 1983: now the regular program in Israel) the treatment of these concepts has been somewhat altered with the aim of improving the students' ability to classify objects as solids or liquids. In addition to teaching the typical properties of solids and liquids, the program explicitly discusses non-rigid solids, powders and dense liquids by raising problems such as 'Is a powder a liquid or solid?', 'Are honey and tahini liquids?' etc., while emphasizing that powder is made up of tiny solid granules and that a dense liquid is a liquid even though it flows extremely slowly.

An evaluative study was undertaken to find out: (a) To what extent may formal instruction improve students' ability to classify the materials and their ability to verbally explain the concepts, 'solid' and 'liquid'? (b) Would formal instruction relating to children's intuitive notions of solid and liquid and dealing directly with the specific problems of classification yield better results than instruction which does not do this?

Students' Definitions of Solid and Liquid

All the seventh grade students demonstrated prior to learning that they could recognize the similarity between liquids and they used the word liquid to describe that similarity. As for solids, about 50 per cent of the students recognized similarity and about 30 per cent of them used the word *solid* to describe that similarity. Following learning, all of the students in both groups (the experimental — *Inside Matter*, and the control — *The Structure of Matter*) could recognize the similarity between solids and 90 per cent of them used the word *solid*. The behavior of grade 9 students and students from teachers college was similar to that of grade 7 students after learning. With regard to defining solids and liquids, we will first discuss those explanations of 'solid' and 'liquid' which related to particles, as they were a direct outcome of what the students studied. It must be pointed out that only about a third of the seventh grade students actually used terms relating to particulate theory after studying and no significant difference between the two groups was observed.

In general, the definitions were incomplete for both solid and liquid, and related usually to only one of their characteristic properties (movement, distance or arrangement of particles). It should be pointed out that in no instance were the students explicitly asked to define the concepts in terms of particles and that they were free to give a physical definition. But even their physical definitions were in many cases incorrect, or failed to include all of the properties of liquids and solids. The majority of students in both groups tended to use one or two properties (hard for solids (40 per cent), spillable or pourable for liquids (35 per cent)) and only a few used as many as four. Since in both study programs the students were taught full physical definitions of the terms *solid* and *liquid*, they might have been expected to know how to use them. Moreover, after having been taught the particulate theory as an explanation for the various states of matter, many more of them might have been expected to use this to explain the concepts, 'liquid' and 'solid'. It is clear, however, that in most cases students continued to regard solids as a hard material in spite of the course, and in spite of the fact that most of them correctly classified most solids and liquids (as will be later shown). The achievements of grade 9 students were similar to those of grade 7 students, in that the former generally used one or two criteria. Teachers college students behaved similarly, except that they tended to use particulate explanations even less, especially for 'solid', though for the most part they named more characteristic properties.

Students' Classifications of Solids and Liquids

In both groups the course resulted in improvement in the ability to classify materials as liquids solids. In the experimental group, *Inside Matter*, there was a rise from 16 per cent in the first test to 60 per cent in the post-test of the number of students who classified all the materials correctly, while in the control group, *The Structure of Matter*, the rise was from 20 per cent in the first test to 48 per cent in the post-test. Of the grade nine students, 48 per cent classified all the materials correctly, while 40 per cent of the teachers' college students did the same.

Significant improvement was observed in the classification of non-rigid solids in both seventh grade classes as a result of their course (experimental group from 48 per cent to 90 per cent, control group from 52 per cent to 70 per cent). The experimental group performed significantly better than the controls. Ninth grade and teachers' college students succeeded in classifying only around 75 per cent of the powders.

Summary

As has been shown above, the teaching of *The Structure of Matter* significantly improves students' ability to recognize similarity among solids, to use the word *solid* to describe that similarity, and to classify materials as solids or liquids. The ability to explain what solids and liquids are also improved but to a much smaller extent. It seems that with regard to the development of the concepts, 'solid' and 'liquid' the classification aspect is relatively susceptible to modification while the learning of formal definition is more resistant to change. Moreover, when the instruction program took into account students' intuitive notions of solid and liquid and dealt with their specific difficulties in classifying materials accordingly, the improvement in classification ability was significantly greater. However, this advantage of the experimental group over the control group in classifying materials as solids and liquids was not paralleled in their ability to explain the nature of solid and liquid. It seems therefore that in spite of the improved classification ability, neither of the programs succeeded in helping students satisfactorily internalize either the physical or the particulate theory explanations of 'solid' and 'liquid'.

Children's Conceptions of Change of State — from Solid to Liquid

Interest in children's understanding of change of state from solid to liquid arises from the findings described above concerning children's ideas about solid and liquid. It is evident from these findings that water serves as an exemplar for liquid, and that children may conceive of liquids in terms of water. However these findings did not allow us to draw a conclusive distinction between two possible explanations of the problem: (a) That the source of this problem is linguistic; i.e., children use the word *water* to describe any liquid, because the word *liquid* is not available to them. This does not imply that they really believe a liquid is water; (b) That the source of the problem is conceptual; i.e., children believe that all liquids are water. We tried to resolve this problem by looking at how well children understood the process of melting an inflammable solid (candle), assuming that if the problem is conceptual, children will relate to the change in state as a variance of substance from candle wax to water, and will regard the liquid wax as an uninflammable substance (water). However, if the problem is linguistic, we can expect children to regard the liquid wax as an inflammable substance. This type of problem is actually a conservation task which deals with the invariance of a qualitative property (inflammability) in the process of transformation by melting. Piaget and Inhelder (1974/1941) referred to such a qualitative invariance as identity, which characterizes preoperational thinking (up to age 7). Thus another question arises: what are the relations between qualitative (identity) and quantitative (weight) conservation. According to Piaget and Inhelder, understanding quantitative conservation is possible only if children are able to understand, in addition to qualitative conservation, also the reversibility of the process, i.e., changing back the liquid wax to its solid state. Therefore, we also tested children on the issue of reversibility.

Identity of Substance

In order to find out whether children grasp the identity of substance during the process of melting a candle, they were directly asked about the identity of substance before and after changing state (i.e., whether the candle before and after melting was made of the same substance). About 50 per cent of the kindergarten children answered correctly and, beginning with the first grade, the majority of children succeeded. Common justifications were: 'It is the same as before, we only heated/dissolved'. Justifications supporting incorrect judgments were: 'It is already water (liquid)'.

Surprisingly enough, when asked indirectly about the identity of substance before and after changing state from solid to liquid (i.e., would the liquid burn if lit), many children responded entirely differently. The general level of success was dramatically lower. Only about 15 per cent of kindergarten children answered correctly, the rate of success in grades 1 to 5 was around 40 per cent and only from grade 6 upwards a steady but slow increase could be observed. Justifications in support of incorrect answers stemmed mainly from the idea that liquid wax is water: 'It is water and will put out the fire'. Some of the children, despite the fact that they responded correctly, thought of the liquid wax as having the properties of water (i.e. wick will burn until it reaches the wax). Others, it seems, believed the wick to be the only substance capable of burning in the candle with the wax serving simply to support the wick, but they did not regard the liquid wax as water. The idea that liquid candle is water and will therefore extinguish the fire is very popular among all age levels and shows that children do not understand the identity of substance in this case. It also supports the findings reported earlier, that children tend to regard all liquids as water. There was an indication that some children in the third and fourth grades were beginning to differentiate various liquids that they previously regarded as water. These students justified their incorrect answers to the inflammability question by saying that 'all liquids except oil (or those liquids which burn fast) put out fire'. Progress in this process of differentiation is reflected in increased success in the inflammability task and justified through the following expressions: 'Both have wax and wick', 'Both are wax', 'It is the same material in different forms'. This process reaches its peak when children's representation of liquid as water includes all liquids, while discriminating between water and other liquids. For example, Renana (seventh grade), when asked to define liquid, said 'It is something watery', but when asked whether liquid wax could burn she answered 'Yes'. Comparison of the responses to the two tasks, which test recognition of the identity of substance, indicates a remarkable difference at all age levels. This may suggest that the direct identity task does not test the identity of substance, but rather relates to the source of the material. Children respond to the question of identification by referring to the source of the material (i.e., what was the material made from).

Conservation of Weight

The second question addressed here concerned the problem of conservation of weight, by using two processes: Breaking a piece of plasticine into small pieces, and melting a candle. In the case of the plasticine, 20 per cent of kindergarten children, and from first grade on practically all children, conserved weight. However, only about 50 per cent of grades 2–5 children conserved weight in the candle problem. While the justifications of the correct answers in both tasks centered around 'It was the same before', 'We just heated/dissolved', 'It is the same material', etc., the justifications for incorrect answers indicate that — except for some kindergarten children who believed solid candle to be lighter — most children assumed liquid wax was lighter than solid candle. Many children believed the solid to be heavier because it was 'harder', 'stronger', etc. Another significant group thought that the candle turned into water which is lighter than the solid, or even weightless. Informal support for these results was obtained with gifted third grade children in a classroom discussion about the weight of objects on the planet Venus. One of the children claimed that 'lead would not have any weight on the planet because the heat would melt it'.

Reversibility

Almost all the children knew that the plasticine pieces could be turned back into a ball. In the case of the reversibility of the melted candle, a gradual increase with age is observed, from 12 per cent in kindergarten to 100 per cent in ninth grade.

But the ability to correctly justify (in addition to the correct answer) lags behind these rates (in about 10–20 per cent difference). Sometimes, children made correct judgments but their justifications were incorrect, i.e., 'It has to be dried or heated in order to turn it back into a solid', which indicates that they perceive of liquid wax as a watery solution. Other justifications for correct answers indicate an understanding of the reversibility of the process: 'By cooling', 'Leave and wait', and these show a gradual increase from 20 per cent in the first grade to 100 per cent in the ninth grade. Children who thought the process to be irreversible explained 'It is impossible to revert liquid (water) into something as hard as stone', or 'It was already used, it is already water'. These children had no problem in understanding reversibility in the case of the plasticine.

Summary

Many children from all age groups tested perceived of liquids in terms of water, and they believed that melted wax is not flammable. They also responded very differently to the conservation of weight tasks for the broken plasticine and the melted candle. These findings seem to be similar to the horizontal decalage reported by Piaget (1955/1937) and Piaget and Inhelder (1974/1941). Our explanation for the decalage observed in our research is that physical knowledge, supported by perceptual aspects of the task itself, interferes with already existing logico-mathematical knowledge. In our case, children who already possessed the logical mathematical knowledge necessary for recognizing the conservation of weight used instead 'incorrect' factual knowledge, verbally expressed as 'Water has no weight because it cannot be held', or 'A solid weighs more because it is heavier or stronger', etc. This factual knowledge is apparently very powerful and is also supported by the immediate perceptual features of the task.

Educational Implications

The studies described above clearly show that the young children up to the age of 5, and the adults with no formal schooling at all, did not understand the concepts, solid and liquid, as they are defined in science. Despite their rich everyday experience with plenty of both solid and liquid substances, neither these young children nor these adults developed the relevant formal concepts, possibly because they are not useful in everyday life.

Children in the educational system, on the other hand, already at kindergarten age, show a significant increase in their ability to recognize similarity between liquids and solids and in their ability to classify objects according to these concepts. It seems that non-specific teaching (teaching of basic thinking skills in the early school years) serves as a catalyst for the development of concepts. Support for this assumption was indicated by the enhancing effect of a teaching program focusing on observation and classification skills. The children started to construct conceptual images of both liquid and solid. The conceptual image of liquid is 'water' or 'something that pours', and that of solid is 'something hard, resistant to change'.

The development of the concepts of solid and liquid is not simultaneous. The concept, 'liquid', because of its perceptual and linguistic components, develops before the concept 'solid'. As children do not conceive of 'solid' and 'liquid' as mutually exclusive, they may create an additional, 'intermediate' group when encountering substances they cannot relate to either of the two categories. This situation remained largely unchanged until the age of 13. At this age (seventh grade — junior high-school), specific teaching related to these concepts enhanced students' understanding of them, improving their classification ability and their ability to define these concepts. Teaching which took account of students' intuitive ideas about solid and liquid, and which related to the specific problems associated with them, led to better achievements in classification but not in definition.

Teaching the physical attributes of the states of matter does not seem to lead to better understanding of the particulate theory (see also Fensham, chapter 2 in this volume). It may very well be that the introduction to this theory should be completely different, for example using dynamic processes such as dissolution through which some students intuitively generate particulate explanations (Inditzky, 1988).

In general, the research findings suggest that teaching should follow the natural sequence and pace of students' learning. As it was shown that children until sixth grade have difficulties in grasping the macroscopic level (the physical attributes) of solids and liquids, it is recommended that they should be taught this level before any attempt to teach the particulate theory of matter. Moreover, enough time should be allowed for the ideas to be incorporated and understood before relating them to the microscopic level.

When teaching the concepts of solid and liquid we should begin with the one for which there exists maximal perceptual reinforcement for correct intuitive knowledge, namely liquid. Only later should we proceed to the one with lesser perceptual reinforcement — solid.

In science instruction it is common to present concepts using typical examples. When the typical examples also have strong supporting perceptual elements additional examples should be looked for. For instance, while teaching the concept 'liquid', the common examples of water, drinks, etc., are usually given. These have strong perceptual elements which help classify them as liquids. If this, however, is not augmented with examples of viscous liquids, inflammable liquids or powders, nothing is added to the existing intuitive knowledge.

It is recommended therefore that the topic, 'states of matter' is taught initially with the focus on the concept 'liquid'. This may begin as early as the first grade with the recognition of similarity between liquids. Among younger age groups, treatment should be on a concrete level, focusing upon similarity recognition, and the correct descriptive use of the word, *liquid*.

Treatment of the concept, 'liquid' should center mainly on properties of different liquids to demonstrate the fact that not all liquids are watery (through demonstration of inflammable and uninflammable liquids, liquids which mix, and others which do not, etc.). Since only seventh grade students begin spontaneously to use examples of non-watery liquids, and since only in the sixth and seventh grades did 50 per cent of children understand that a change in the state of matter does not convert a solid to water, we can assume that this type of treatment of the concept, 'liquid' can probably begin in the upper grades of elementary school (fourth-sixth grades).

In teaching solids, special treatment should be given to the sub-groups of non-rigid solids and powders, accompanied by examples which demonstrate the reasons why they are solids. Up to the third grade, children usually do not spontaneously recognize similarity between different solids. The use of the word, *solid*, appears only in the fifth grade. Therefore, it is recommended that teaching of rigid solids should not begin before the third grade. Teaching should then be on a concrete level only, focusing upon the recognition of similarity between rigid solids, and the use of the word *solid* to describe them. Treatment of non rigid solids should start only after the children have apprehended the concept of rigid solids. To their primary understanding that a solid is a hard substance which does not change shape, can by then added the condition that it does not change as long as strong enough force has not been applied to it. Only at this point can powders be treated as solids. Up till fifth-sixth grade children's intuition leads them strongly to assume that powders are liquids. Only at this age, therefore, is it possible to draw children's attention to the fact that a powder is a small-grained mode of solid, and hence is a solid. As long as the child believes that powder is liquid, and we supply it with a formal definition of solid in which 'shape and volume are fixed', we reinforce the incorrect concept of powder. If explanations of the state of matter based on the particulate theory are taught to students with this incorrect concept, it is likely that they will infer that powder grains are the particles to which the particulate theory refers.

With regard to the specific problems which this chapter aimed to address, it was found that young children do not distinguish between the concepts, 'solid', 'hard', 'heavy' and 'strong'. In order to overcome this difficulty it is possible to present them with substances which are characterized by one salient property such as substance which is 'heavy' but not 'solid' (for example — mercury), or a substance which is 'hard' but not 'heavy' (for example —aluminium), etc.

Treatment of the idea according to which powders are liquids can also consist of demonstrating similarities and differences between powders and liquids: both powders and liquids pour and assume the shape of their container, while it is possible to pile up powders but not liquids; liquids make drops while powders do not.

Since many children have problems conserving weight in the process of a change of state and in recognizing the reversibility of the process, it is worthwhile enabling them to experience the phenomenon by weighing the materials before and after a change of state.

Notes

1 This chapter is based on the book: *Solid is something hard — The Concepts Solid and Liquid: Development and Learning* by Ruth Stavy and Dina Stachel, Ramot Publishing Co., Tel Aviv University, Israel, 1986 (in Hebrew).

2 For the classifying experiment we used the following substances: The liquids included thin liquids (water, chlorinated water, spirit, corn oil, gasoline), and dense liquids (honey, chocolate syrup, hair shampoo, mercury). The solids included rigid items (coin, metal cube, ice, chalk, chocolate, glass, candle), non-rigid items (cloth, cotton-wool, dough, plasticine, sponge, aluminium-foil, steel-wool, metal spring), and powders (aluminium powder, sawdust, flour, sand, sugar).

References

BOURNE, L.E., DOMINOWSKI, R.L. and LOFTUS, E.F. (1979). *Cognitive Processes*. New Jersey: Prentice-Hall.

DICKENSON, K.D. (1982). *The Development of Children's Understanding of Materials*. PhD dissertation, Harvard University.

DRIVER, R. *et al.* (1987). *Approaches to Teaching the Particulate Theory of Matter*. Children's Learning in Science Project. Leeds, UK: Centre for Studies in Science and Mathematics Education.

INDITZKY, R. (1988). *The Development of Understanding the Particulate Nature of Matter Among 7–13 Year Old Children*. MA Thesis, School of Education, Tel Aviv University, Tel Aviv, Israel.

ORPAZ, N. (1983). *Inside Matter*. (experimental edition). TL (in Hebrew).

ORPAZ, N. and BEN-ZVI, R. (undated). *The Structure of Matter*. TL (in Hebrew).

PIAGET, J. (1955/1937). *The Construction of Reality in the Child*. London: Routledge and Kegan Paul.

PIAGET, J. and INHELDER, B. (1974/1941). *The Child's Construction of Quantity*. London: Routledge and Kegan Paul.

STACHEL, D. and STAVY, R. (1986). *The Effect of Teaching on the Understanding of the Concepts 'Solid' and 'Liquid' by Kindergarten Children*. (ERIC Document Reproduction Service No. ED 291 557).

STACHEL, D. (1986). *MATAL Early Childhood Program*. Tel Aviv: Masada. (Hebrew)

STAVY, R. and STACHEL, D. (1985a). Children's ideas about 'solid' and 'liquid'. *European Journal of Science Education*, *7*, pp. 407–21.

STAVY, R. and STACHEL, D. (1985b). Children's conception of changes in the state of matter: From Solid to Liquid. *Archives de la Psychologie*, *3*, pp. 331–44.

STAVY, R. (1987). '*The effect of instructional programs on the understanding of the concepts "solid" and "liquid" by junior high school students*'. Unpublished manuscript. School of Education, Tel Aviv University.

STAVY, R. (1991). Using analogy to overcome misconceptions about conservation of matter. *Journal of Research in Science Teaching*, *28*, pp. 305–13.

Chapter 17

Pedagogical Outcomes Of Research In Science Education: Examples In Mechanics And Thermodynamics

Laurence Viennot and S. Rozier

A considerable body of research findings is now available concerning pupils' and students' own ideas in science. The corresponding difficulties for teaching specific contents have been now widely identified. A question of major significance is therefore to see to what extent this knowledge provides some pieces of information about what we should do in teaching.

This chapter concentrates on the so called 'pedagogical implications' of research about learners' ideas in physics. Four categories of such implications will be described and exemplified.

Awareness

If teaching is to take into account learners' own ways of thinking, the first thing to do is to provoke awareness in teachers about their students' ideas. Most probably, reading research papers in this field is not enough, and teachers should be able to check by themselves what was found by others. These situations are all the better when they are easy to duplicate. Moreover, in order to avoid *ad hoc* ways of interpreting learners' responses, such situations should be grouped into sets focusing each on a given feature of reasoning through different physical contexts. Such sets of situations are now available. The question of how to use them to provoke learners' awareness about their own ideas is also of major interest.

Vigilance

It has been very frequently observed that pupils' or students' 'naive ideas' are, in fact, widely shared. In particular, numerous echoes of such conceptions can be found in currently available textbooks, and *a fortiori* in popular papers and other media. Therefore, there is a kind of permanent reinforcement of corresponding ideas, or at least of ways of seeing things, in our students. Teachers need active vigilance in that respect, if our students are not to continuously replace scientific ideas by common ones. This may appear in several forms: identifying clearly incorrect statements, or, more subtly, localizing ambiguous comments or schemas which are more or less compatible with common ideas. Students can be proposed such activities as text criticism, with appropriate help. This refers in fact to the following item.

Guidelines for Teaching

A sound analysis of a given conceptual domain and of corresponding learners' ideas may provide some indications on how to teach it. The basic hypothesis here is that we should take these ideas into account in the (supposedly) best possible way, and design our teaching strategies accordingly. It is not argued, indeed, that provoking awareness is enough to change learners' ideas.

Discussing Our Teaching Goals

This is ultimately the most crucial 'pedagogical outcome' of studies of students' ideas. Indeed these show very crudely the fact that our present teaching goals do not take into account some points that actually raise severe difficulties for our students. So the question is asked: what do we want our students to understand?

These levels of pedagogical outcomes will be exemplified and discussed with two different kinds of research:

- a content-specific topic: the problem of differentiating the second and third laws of Newton, and not mistaking a balance of forces acting on a given body for the comparison of the two forces implied in a given interaction;
- a 'transverse' topic: the difficulties of coping with multivariable problems.

The four types of pedagogical implications are, in fact, interdependent. They will be dealt with in the most appropriate order, which is different for each topic as explained in the final remarks.

A Content-Specific Field of Research: Learner's Ideas in Mechanics

Common Ideas About Second and Third Laws

Did Newton find two different laws? Such is the question that could be asked when considering pupils' or students' comments in elementary dynamics. Indeed, students' ideas about the second and third laws seem deeply intricated.

The difficulties raised by the second law (total force exerted on a body A = mass of A × acceleration of A) were among the first ones extensively explored by researchers (Viennot, 1979a, b; Clement, 1982, McDermott, 1984). The main features of students' ideas on this topic, and corresponding common comments, are the following:

- Notions of 'force', 'velocity' (if not simply 'motion') seem either not differentiated or quasi proportional:
  'Velocity is null, therefore force is null.'
  'Forces are different since motions are different'.
- 'Force', 'energy', 'impetus', 'inertia', are not clearly differentiated:
- 'Force' is ascribed to the moving body:
  'The upward force of the mass'.
  *'The mass **has** some upward force, otherwise why would it hold in the air at the top of the trajectory?'*

These features do not appear necessarily all together, but they are so often associated at least two by two that this suggests the existence of an underlying reasoning. This will be referred to as 'V-F' reasoning in what follows. The anthropomorphic connotation of this reasoning is quite manifest.

What about the third law?

This law, contrary to the second one, puts into play two distinct objects, say *A* and *B*. The well-known relationship, $F_{A\ on\ B} = -F_{B\ on\ A}$, holds whatever the motions of *A* and *B* may be. For example, when a mass suspended from a spring is accelerating upwards, the action of the mass on the spring is still exactly opposite to the reaction of the spring on the mass.

But an anthropomorphic view of such a situation leads to seeing it as a dynamical conflict between two objects in which the strongest of them determines a global motion in the direction of its own effort (Viennot, 1979a). Numerous comments suggest this interpretation: '*the force that the mass opposes to the spring*', or '*the spring imposes its force to the mass*'.

The 'V-F' reasoning applies then with an 'F' which is the 'total' force in a 'balance' implying two distinct objects. A common conclusion is that: '*The reaction of the spring is larger than the action of the mass*' or '*The third law does not hold any more*'.

So, common ideas about Newton's laws show a lack of differentiation between these laws.

Pedagogical Outcomes

(a) Awareness

On this topic, it is extremely easy to uncover evidence of common ideas that are not consistent with Newton's statement of his third law. It is enough to put *motion*, and preferably *contact* in the proposed situation. If the interaction puts into play at least one animated body, the common ideas described above appear all the more frequently, but this condition is not necessary.

Thus, a driver (*A*) pushing his car (*B*) on a flat road towards a garage, a traveller (*A*) in an aircraft, lifting his luggage (*B*) up into a locker, a nail (*A*) hammered into a board (*B*), a mass (*A*) suspended from a spring (*B*) considered a very short time after having been released, such are some of the situations about which the question '*Is the force exerted by A on B larger than the force exerted by B on A?*' is extremely, and easily, instructive.

(b) Guidelines for teaching

We have seen that one of the facets of learners' difficulties is to mistake forces acting on distinct objects for forces acting on the same object.

We pointed out above that, in students' comments, a force on a mass *A* easily becomes 'the force of the mass'. But the force exerted by *A* on *B* is most commonly referred to as the force of mass *A* (on *B*). This verbal closeness paves the way for mistakes about what is acting on what. For instance, the weight of an object is often said to be the force of this object on its support.

Further, the usual way of representing forces, for instance for a mass suspended from a spring and accelerating upwards (see figure 17.1, schema a), also favours such a mistake because it does not specify clearly enough which objects the forces are acting on: is the weight of the mass also acting on the spring? Often, it seems as if forces were acting on the contact point between the two objects, a way of maximizing the indetermination. Such an analysis suggests that one should draw the forces in a more explicit way, objects being represented with a space between them even if they are actually in contact (figure 17.1, schema b). A different coding for each interaction, the compulsory mentioning of *two* forces per interaction — gravitation included — are additional aids which require students to perceive, for instance, that mass M does not 'exert its weight on the spring', or that Archimedes' interaction also results in a downwards push of the immersed body on the liquid. In such a mapping, equality between intensities of two interactions — at rest — does not mean an indistinction (see schema c on figure 17.1). The weight of the mass is explicitly 'married' with the opposite force exerted on the earth; and its action on the spring, though of the same intensity, is 'married' with another force, the force exerted by the spring on the mass.

From this point of view, it seems preferable to introduce interactions at a distance *before* interactions of contact. This may seem paradoxical. An interaction of contact, indeed, seems more familiar and accessible to students than an interaction at a distance. But this is a misleading easiness, as shown above. A correct analysis is in fact easier for interactions at a distance, because drawing the objects with a space between them is then only representing reality, and does not require any further effort of abstraction.

(c) Vigilance

One can expect to find in written materials some comments analogous to those quoted above. However, it is important to distinguish several levels of compatibility between common ideas and what can be read in textbooks. Most probably, indeed, our vigilance should not be limited to denouncing errors: other features of pedagogical tools, which are at first sight of no consequence, may have determining effects, in reinforcing or at least in making comfortable common erroneous ideas.

The most remarkable manifestation of common ideas in textbooks is that of bluntly erroneous statements. Thus, one can find this comment in a French textbook for grade 9 (Lacourt, 1981):

> If a force **F** is exerted by a nail on a board, this simultaneously exerts a force –**F** on the nail. If force **F** is increased, reaction –**F** increases as much. Reaction and Action balance each other. If the increase in **F** is too large,

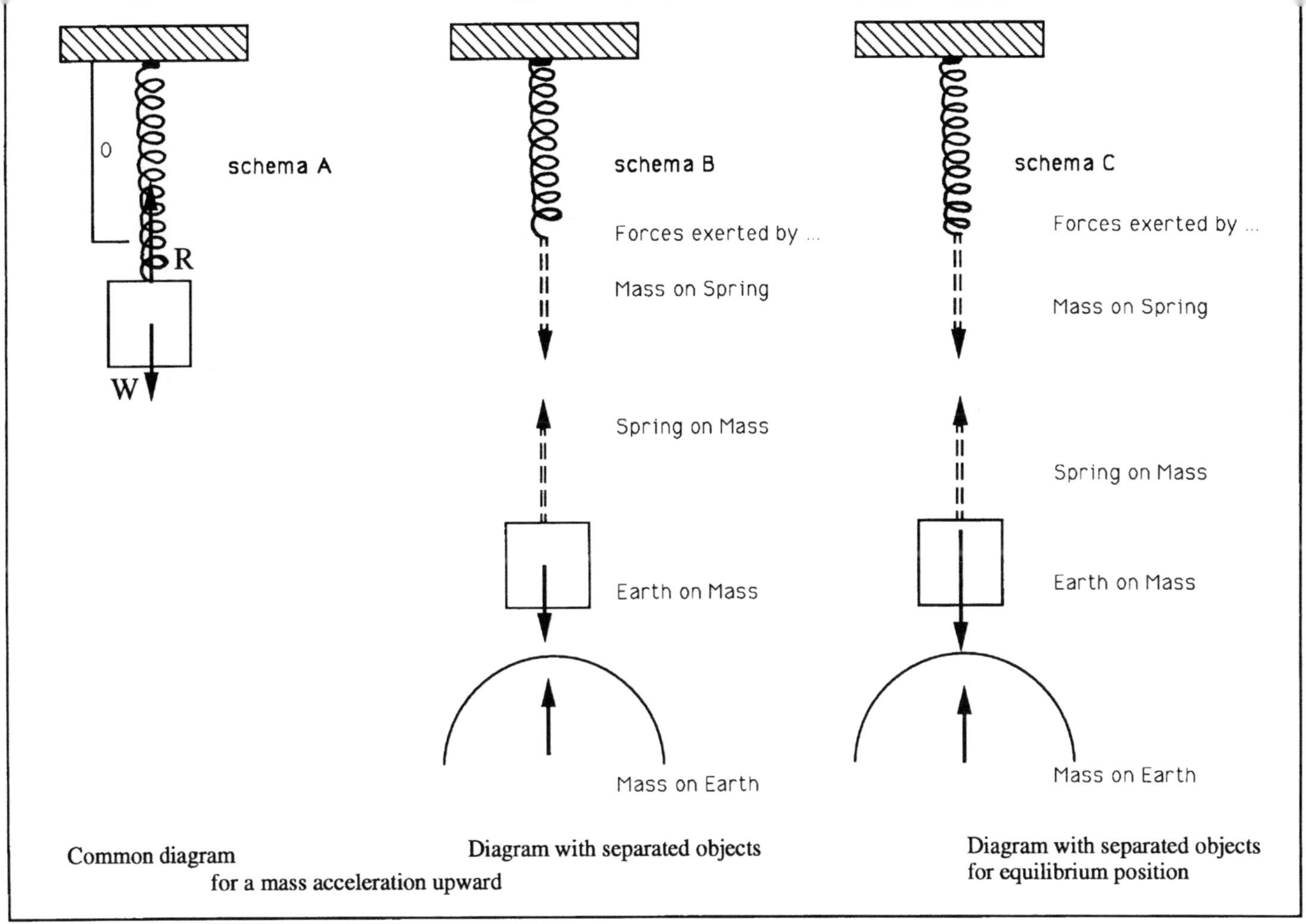

Figure 17.1: Two different ways of representing forces (a: common; b: and c: with separated objects)

> then Reaction cannot any longer balance Action: the nail penetrates into the board!

Despite some papers published subsequently about this problem (Viennot 1982 and 1989), a later textbook (Fontaine and Tomasino, 1990) claims:

> Consider the simple case of a brick settled on the table. The brick exerts on the table a force $\mathbf{F}_{B,T}$ that is clearly identical to its weight **P**. The table exerts on the brick a force $\mathbf{F}_{T,B}$ called reaction **R** of the table on the brick . . . Action and Reaction have the same intensity as long as the table is strong enough. If weight **P** is too large, the table will not be any longer able to ensure the equality $\mathbf{R} = \mathbf{P}$, and it will collapse.

This indicates, if there was any need for it, the enormous resistance of such common ideas to teaching (textbook writers are rarely autodidacts), and to mere coherence. Indeed the last comment, for example, is contradicted one page further in the same book, by this statement:

> The principle of Action and Reaction holds for any interaction between two solids, be they in motion or not, in contact or not.

More subtly, one can notice the non-neutral character of a comment such as: '*A mass is suspended from a spring. The mass exerts its weight on the spring, that lengthens*' (Chatroux, 1981). According to their teaching style and mood, teachers diversely estimate the correctness of this statement; most of them do not even see the least problem about it. However, it obviously goes hand in hand with the common and plainly erroneous comments quoted above, and in the authors' view, it *is* erroneous. Indeed, the action of the mass *before* the spring has lengthened is said to be '*its weight*', while it is, in fact, as small as the quantity '*kx*' (with usual notations) at the beginning of the extension. At the least, this comment is ambiguous.

What to say, finally, about a schema analogous to figure 17.2 (Saison, 1981)? Very common and certainly not incorrect, it however 'maximizes the indetermination', as explained above. It is likely to reinforce the ideas described above, and to favour a total identification of two interactions, that (gravitational) with the earth and that (of contact) with the spring.

(d) Discussing teaching goals

At this point, the reader may feel quite uncomfortable, and think that physics is really too demanding. Incidentally, are 'errors' so clearly errors? Can we afford an intermediate level of 'correctness'?

Here, we undoubtedly get to a debate about teaching objectives.

If our goal is to teach the third law, statements contradicting this law must certainly be said to be 'errors'. All comments in which a comparison of reciprocal forces is based on the fact that there is, or there is not, a motion somewhere are erroneous, i.e. contradictory with what we would like our students to understand.

If not, the question is posed of what else we can define as an intermediate conceptual level concerning this law. For many topics in physics, such a specification is at the same time possible and unavoidable. A given domain can then be approached through successive 'layers' of increasing conceptual complexity, each one being at the same time self-consistent and compatible with the essential

Figure 17.2: A schema for a mass suspended from a spring

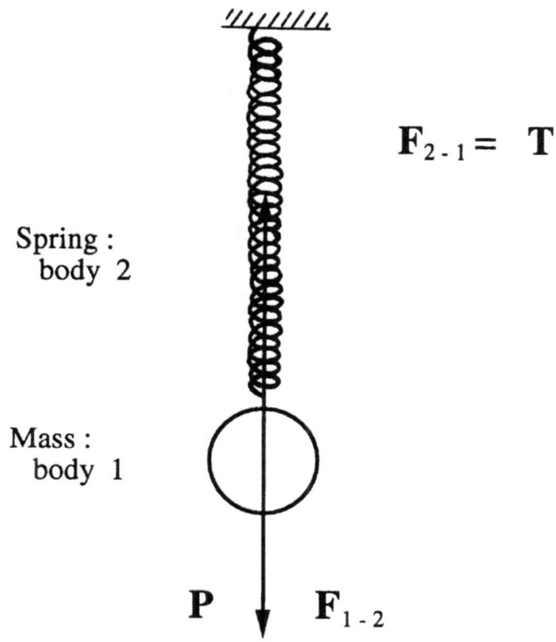

features of the concepts and laws at stake. In the case of the third law, it is not obvious how this can be fruitfully done. In particular, illustrating the third law only in situations of rest, even if one says that its validity is not restricted to such situations, is of quite questionable interest. In other words, the first valuable 'layer' is already very high.

The choice made up to now in France, although implicitly, was not to teach how to discriminate the two Newton's laws, and therefore not to teach the third one. Indeed, teaching of elementary dynamics is focused on mechanics of point, or at most, single, bodies, at the expense of a systemic view. The alpha and omega of such teaching is the second law. In the author's view, the third law is then taught just for saying it has been taught.

A Transverse Problem: Coping with Multivariable Problems

Let us now examine research results about a question that is not restricted to a given chapter in physics: how to cope with multivariable problems? In fact, this question is omnipresent in physics.

What 'Should' Be Understood

When several variables are necessary to specify the state of a system, analyzing its transformations requires understanding the following points:

- During the transformation, most of the time, several variables change *simultaneously*.
- Some relationships between these variables hold *permanently*.

In order to reduce the number of variables which are necessary to specify the state of a system, and to get simple relationships between them, it is sometimes necessary to introduce some hypothesis such as the quasi-static character of a transformation in thermodynamics or of phenomena in electric circuits. The basic idea in this way of modelling systems is that a kind of global equilibrium is present at any time during the transformation. This permits us, for example, to define a pressure for a gas which undergoes a quasi-static transformation, or to have the same intensity of current in every part of a series circuit. Writing $F = -kx$ for an oscillating spring in fact is based on the same idea.

How It Is Understood

(a) A general trend towards 'functional reduction'

Research findings (Viennot, 1985; Rozier and Viennot, 1991; Maurines, 1992) show that all this gives rise to considerable difficulties.

A first common tendency is to take into account fewer variables than necessary, and preferably only one. This trend towards a 'functional reduction' is a quite general and well-known fact. But it is interesting to see in more detail how this more or less surreptitiously manifests itself in physics.

(b) Combining variables into a single notion

One of the most pregnant ideas in mechanics, as said above, is an adherence between notions of velocity and force. These two quantities are then combined into a single one, half dynamical, half energetical, that might be called a 'supply of force' (Viennot, 1979a).

The same kind of adherence can be observed concerning height and speed for a bump that propagates on a rope (Maurines, 1992). Students often consider these two quantities to be linked, as facets of a unique notion that might be a kind of 'strength' of the bump.

In the same way, 'thermal motion' is often thought of as an aggregate of two quantities, mean distance between particles and mean speed of particles, which in fact are not directly nor necessarily linked (Rozier, 1987; Rozier and Viennot, 1991). At the same temperature, indeed, all bodies have the same mean speed of particles, whatever their state — gas, liquid or solid — may be.

To sum up, one of the ways often used to reduce the number of variables actually taken into account in a complex problem is to combine them two by two. Such couples are then understood as facets of a unique underlying notion.

Another simplifying procedure is the following.

(c) Linear reasoning: an example in thermodynamics

An example from the study by Rozier (1987; see also Rozier and Viennot, 1991) illustrates the 'linear' aspect of common reasoning about multivariable problems. In this reasoning, all the necessary variables are put into play but, in the explicative chain, each phenomenon is specified with only one variable.

Figure 17.3: Questions about an adiabatic compression and correct and typical responses

QUESTION :

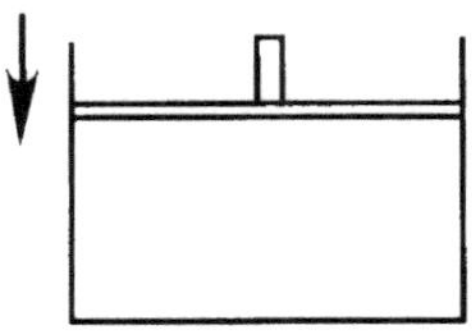

An adiabatic compression of a perfect gas: pressure and temperature both increase. Can you explain why in terms of particles?

<u>notations used below</u>: volume of gas: V, number of particles per unit volume: n, pressure of gas: p, temperature of gas: T, mean speed of prticles: v, mean kinetic energy of particles: ec, heat : Q $\nearrow$: "increases", $\searrow$:decreases", $\lrcorner\!\uparrow$: "is produced", $\rightarrow$:therefore" (see text)

P outlines of .....
.... correct explanation:

$V\searrow \rightarrow \begin{bmatrix} n\nearrow \\ \text{and} \\ v\nearrow \end{bmatrix} \rightarrow \begin{bmatrix} \text{number of collisions}\nearrow \\ \text{per ....} \\ \text{and } v\nearrow \end{bmatrix} \rightarrow p\nearrow$

........common explaination :

$V\searrow \rightarrow n\nearrow \rightarrow$ number of collisions $\nearrow \rightarrow p\nearrow$

T outlines of ......
..... correct explanation :

$V\searrow \rightarrow v\nearrow \rightarrow ec\nearrow \rightarrow T\nearrow$

common explanation :

$V\searrow \rightarrow$ number of collisions $\nearrow \rightarrow Q\ \lrcorner\!\uparrow \rightarrow T\nearrow$

Figure 17.3 shows an outline of a question posed to students (N = 111, first four years of higher education with similar results per year). Commenting on adiabatic compression of a perfect gas, half of the students gave answers such as:

> The volume decreases, therefore the molecules are closer to one another, therefore there are more collisions and the pressure increases.

This common answer can be outlined in the following way: $V\searrow \rightarrow n\nearrow \rightarrow p\nearrow$. It shows a linear shape, i.e. each implication links a single variable to a single other one, contrary to what is needed in correct answer (see figure 17.3). In

Figure 17.4: A question about an isobaric heating of a gas, and correct and typical responses

QUESTION :

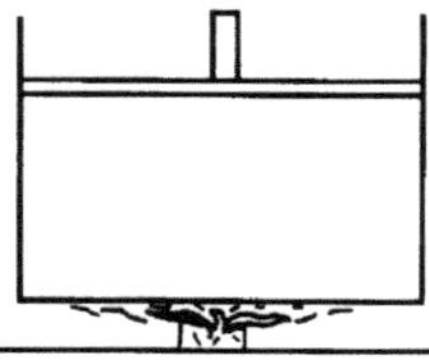

A perfect gas is heated at constant pressure.....
It's volume and temperature both increase.

Can you explain why?

notations used below : see figure 17.3, and C_p : molar specific heat at constant pressure, R: constant, N: total number of moles, Δ algebraic increment of ....

Outlines of ......
correct explanation:

Q (supplied to gas) = $C_p\ \Delta T$
and $Q > 0$
and $C_p > 0$ → $\Delta T>0$
and
$PV = NRT$
and
p,N,R, all constant → $\Delta V>0$

........common explaination

Supply of Q → T ↗ → P ↗ → V ↗

particular, it shows a 'preferential association' of pressure (p) with mean density of particles (n), while p in fact depends on n *and* on mean speed of particles, $\bar{v}$; should $\bar{v}$ be lowered in a transformation (the gas being cooled), n might well increase and p decrease at the same time.

(d) Story-like arguments

Another important feature of common reasoning that Rozier pointed out is a subtle input of time in arguments which, in a quasi-static analysis, should be on a purely logical ground. Figure 17.4 shows an outline of another question posed to the same type of students (N = 120) about an isobaric heating of a perfect gas. A common argument (≈ 40 per cent) to explain the increase in volume of the gas can be sketched as follows:

'supply of heat → T ↗ → p ↗ → V ↗' (T and V are respectively the temperature and volume of the gas).

Again a linear shape is observed. But it is also quite remarkable that the statement 'p ↗' contradicts the data ('isobaric heating'). Rozier's analysis shows that this is probably not the case for students who use this argument. Indeed some of them explicitly describe the transformation in two steps:

- First step: supply of heat → T ↗ → p ↗, piston being blocked
- Second step: p ↗ → V ↗, the piston is released and moves until p equals the external pressure

The status of the explicative chain is then that of a story, which rubs out the contradiction just pointed out, as the higher pressure is then seen only as a temporary phenomenon.

It is worth noting that the horizontal arrow, in such an argument, does not mean only 'therefore', but also 'later'. These logical and chronological levels also melt in the verbal equivalent of an arrow, the word 'then'. Incidentally, such totally ambivalent words are also present in many other languages, for example 'alors' in French, 'entonçes' in Spanish, 'allora' in Italian.

(e) Linear and story-like arguments: a transverse trend

Another example, from mechanics, illustrates the fact that, functional reduction and input of time are not features of reasoning restricted to thermodynamics. In his study of problem solving strategies in physics, Fauconnet (1981 and 1984) proposed the situation and question outlined in figure 17.5 to students at the end of secondary school. It deals with the extension of two springs suspended end to end from the ceiling, the lower end being pulled downwards. A typical comment, also quoted and analyzed in figure 17.5, shows features which are analogous to the ones described above:

- A local analysis, which focuses first on the lower end: the lengthening of the lower spring is calculated as if the intermediate point was motionless, in other words a single displacement is taken into account.
- A story-like argument: the higher spring is said to extend a certain time *after the lower one*, under the action of the transmitted external force. Needless to say, the total extension calculated in this way is smaller than the displacement of the intermediate point!

In this second example, the spatial extension of the system favours an occurrence of linear causal reasoning. This is also the case for what has been called 'the sequential reasoning' (Closset, 1985), or the 'sequential model' (Shipstone, 1984) in electricity. But sometimes there is no spatio-temporal basis to support a sequential analysis and yet a linear causal reasoning is still observed. For instance in mechanics, a study by Menigaux (1991 and 1992) shows that students at the end of secondary school or beginning university studies are reluctant to envisage *simultaneous* motions of translation and rotation, and prefer, at least in the situation proposed in this study (see figure 17.6), to think that they occur *successively*. Many other examples can be found in other domains, for instance in economics (Rozier, 1987).

Figure 17.5: A typical response for a classical exercise, and Fauconnet's analysis

QUESTION:

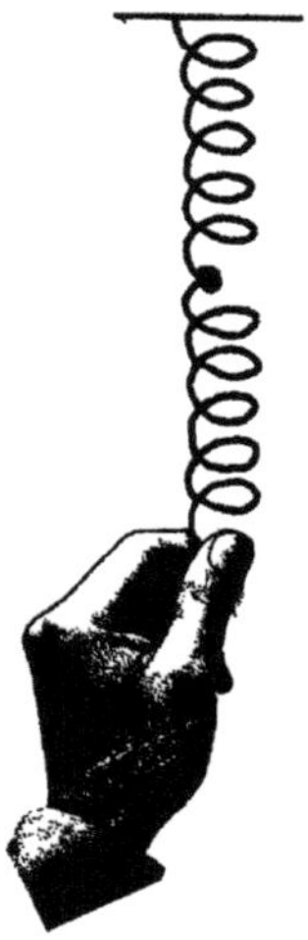

One pulls on the lower end and displaces it by 10cm. What force is it necessary to exert and what is the displacement of the intermediate point?

Notations and numerical values used below:
K_1, K_2: spring constants (resp. lower and upper spring)
K_1=3N/cm, K2=2N/cm
a_1, a_2, a: extensions of, resp., lower, upper, "total", springs.

A TYPICAL RESPONSE	**FAUCONNET'S COMMENT**
$\mathbf{F}$ external = $\mathbf{F}_1$ (lower spring)= $\mathbf{K}_1$* a_1 = 3 *10 = 30N	The extension of spring 1 is mistaken for the displacement of lower end: a local analysis
Force $\mathbf{F}_1$ is then transmitted to spring 2 30N = K_2X $\mathbf{a}_2$ $\mathbf{a}_2$ = 15cm	... a temporal shade ... (this is larger than the total extension!)
The first spring should extend and the other one after a certain time, should do the same	.... explicitly confirmed

Figure 17.6: A situation proposed by Menigaux

SITUATION: a puck on a horizontal table; a constant force is applied always at the same point

QUESTION: will the puck take the same time, a longer or a shorter time to hit the edge PQ when the constant force is applied always at point A or at point B?

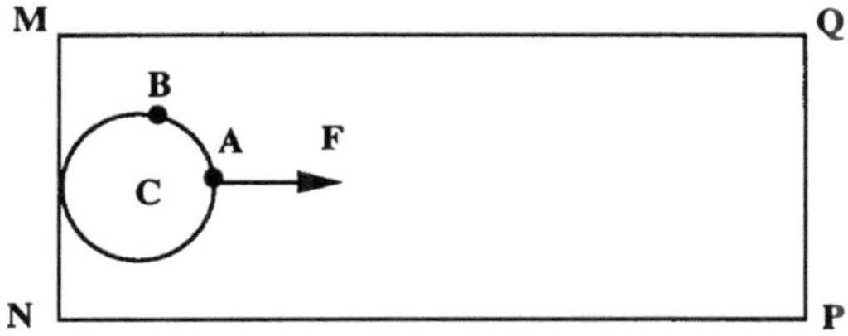

Pedagogical Outcomes

(a) Awareness

The first level of pedagogical outcomes of such research findings is, as usual, a better awareness about corresponding difficulties. In fact this is not always very easy to provoke because, most of the time, the statements relying on linear causal reasoning are not clearly erroneous. Often, they just need some specifications to become correct. For instance, saying $n \nearrow \longrightarrow p \nearrow$ for an adiabatic compression is mentioning two real phenomena, but the first one can be considered as a cause for the second only when the mean velocity $\bar{v}$, or equivalently the temperature T, is kept constant.

So, to make obvious the existence and limits of linear causal reasoning, it is necessary to put in evidence an inconsistency in a given argument or a contradiction between two arguments of this type.

An example of such an inconsistency is given in figure 17.4. The question used is rather propitious to an awareness about linear causal reasoning and it shows rather clearly the incoherence to which such a way of reasoning leads.

It is also possible to uncover the specific features of students' reasoning by asking whether their arguments are reversible. Confronted with a comment such as '*volume V decreases then pressure p (inside the gas) increases*', the teacher may ask how it is that the reverse statement '*pressure p increases (still inside the gas), then volume V decreases*' seems so surprising. If this statement were only referred to a one-to-one relationship, such as $F = -kx$ for a given spring, both implications would be equally acceptable. But, in fact, each of them is understood as a particular story. The first story is familiar, and goes implicitly with a (sufficiently) constant temperature. In the second one, a cause is lacking to explain the decrease in volume, because what is said before cannot be understood as a cause.

(b) Vigilance

An excerpt from a book of popular science (Maury, 1989) illustrates the kind of *ad hoc* variations on a multivariable problem introduced by the common tendency towards 'functional reduction':

> Planes fly very high, at an altitude where molecules of air are much less numerous, and therefore the pressure of the external air on the window is much lower than at sea level.

This explanation may be summed up in the following way: n ↘ ⟶ p ↘, nothing being said about temperature. Incidentally, the temperature is much lower at the altitude considered (≃ 10km) than at sea level (≃ 70° K, i.e. a decrease of about 25 per cent in temperature) which also contributes to the lowering of pressure. Five pages further in the same book, the hot air balloon is presented and 'explained' using the fact that when the temperature increases, it contains 'less and less air'. So the 'number of particles . . .' decreases. Following the previous argument about planes, the pressure inside the hot air balloon would be lower than outside. This is not the case. No connection is made in the book between the two proposed explanations. The point here is not to provoke indignation, but to use such texts for an in depth analysis of the limits of (over)simplified 'explanations' based on functional reduction.

Another example illustrates that vigilance can also bear on more subtle features of texts. The following excerpt from a textbook has been proposed by Rozier (1987; Rozier and Viennot, 1991) to university students, who were asked to read it carefully:

> Thermal energy possessed by each molecule is large enough to prevent the molecules of the gas from being bound: in a gas, molecules are continuously hitting each other and bouncing. But if temperature is lowered, the system will be able to become liquid and even solid. Such physical phenomena occur when, with decreasing temperature, molecules have so low a mean kinetic energy that they cannot any longer resist the electromagnetic interaction. They first gather in liquid state and finally get bound in solid states.

The subsequent questions are:

1 Do you think that this text suggests the following statements:
Statement No. 1: At a given time during the liquefaction, mean kinetic energy of a molecule of gas is larger than mean kinetic energy of a molecule of liquid (liquid and vapor are in thermal equilibrium at the time considered).
Statement No. 2: At a given time during the liquefaction, the mean distance between particles is larger in the gas than in the liquid.

2 Do you think that

Statement 1 is	true	false	why?
Statement 2 is	true	false	why?

Among 181 students in the three first years at University, 77 per cent think that the text suggests statement no. 1 and 69 per cent think that this statement is true.

(The corresponding percentages for statement no. 2 are 80 per cent ('the text suggests statement no. 2') and 85 per cent ('statement 2 is true')).

As said before, mean kinetic energy (ke) depends only, in classical thermodynamics, on temperature (T) and is therefore the same for two phases of a substance at thermal equilibrium. This is recalled one page further in the book. How is it that so many students have 'read' something that the author did not intend to write?

Apart from the fact that one often sees in a text what one already thinks (see the rates of responses to question 2), this surprising finding can be interpreted in the following way. There is a strong input of temporal connotation in the text: 'If . . . the system will be . . . , . . . they cannot resist any longer, . . . first, ... finally'. This gives a story-like structure to the argument, which can be outlined as follows:

$T \searrow \longrightarrow ke \searrow \longrightarrow$ electromagnetic interactions win $\longrightarrow$ liquid state $\longrightarrow$ solid state

The story begins with gaseous state and finishes with liquid state. This prevents students from envisaging simultaneously the two phases at same temperature, and leads them to understand, as they say, that mean kinetic energy (often called 'thermal motion') is larger in the gas than in the liquid. Nothing is clearly incorrect in this text, but it is written in perfect resonance with linear causal reasoning. Hence, it in fact carries an important hidden — and erroneous — content.

So, as observed about mechanics, different levels of subtlety may be envisaged in an effort towards a better vigilance. But thermodynamics is still more propitious to falsely simple arguments. Mastering these requires a high competency on the teachers' behalf.

(c) Guidelines for teaching

Again the question might be posed of what is 'allowed' in teaching in regard to 'correctness' of arguments. Is it possible to avoid being totally paralysed, given the complexity of phenomena? Are there some intermediate ways of explaining physical phenomena that would be at the same time self-consistent and teachable to beginners in physics? For instance, to what extent should we let functional reduction and linear causal reasoning run for a first easy access to physics?

It seems coherent, after considering research findings, to propose as guidelines for teaching:

- *to be extremely careful about the degree of 'explanation' actually aimed at*, and to specify what cannot be accounted for in the frame of the proposed description. Thus, for instance, the following levels of understanding may be envisaged:

  > Gases can change their volume to a large extent because they are made of particles that are not in contact, but (without the beginning of a kinetic theory) we cannot explain why they resist a compression *before* particles are in contact.
  >
  > Solids expand when heated (or contract when cooled), we cannot (yet) explain why. Knowing that thermal motion increases (or

decreases) in such a case is not enough to explain why this makes the solid expand (or contracts). Indeed, the particles might vibrate more intensely, and stay around the same place without drifting (if the potential of interaction between particles was harmonic).

At equilibrium between, say, liquid and gas, thermal motion (mean kinetic energy) is the same in the two phases, and we cannot (yet) explain this surprising thing. In other words, we cannot explain why, with the same thermal motion, some molecules are linked to each other and others are free. We cannot explain why thermal motion keeps the same during the change of state. We know indeed that an input of heat is used to break the links between particles in the liquid. But we do not know why it is used only for this and not also to increase thermal motion.

- When working with some *'soft' explanations, do not hide the dangers of a careless extension* of such explanations to other cases. For instance:

At an altitude, there are fewer molecules, therefore pressure is lower . . . adding: 'this reasoning works only if the molecules have (more or less) the same velocity in the two compared cases'.

When a tyre is heated up, it becomes harder because the molecules have a larger mean speed . . . adding: 'this reasoning works only if the same number of molecules is still in the same volume' (obviously not the case since the tyre is harder, but an approximate constancy of volume may be invoked).

These ways of making *qualitative reasoning hard*(er) may be considered too demanding, but are the price to pay for consistency in such domains.

- Not to hide the non-trivial character of the *permanent validity of laws along time*, during a transformation. This is implicitly considered as obvious, but, as seen above, it is especially difficult to stick to this idea in analyzing complex systems or motions. It might be useful to write fundamental laws or phenomenological relationships in a way that, at least on some occasions, would make explicit this point: $\mathbf{F}(t) = \mathbf{m}\,a$ (same t), $\mathbf{F}(t) = -\mathbf{kx}$ (same t), etc . . . The kind of exercises outlined in figure 17.5 might also be useful to show that this imposes some strong constraints in reasoning.
- Carefully *balance* all the teaching procedures putting into play *decomposition*, step by step analysis, simplified situations, etc., by some episodes where *combined phenomena* or motions simultaneously occur. This in order to consider what can still, or cannot any longer, be said.

(d) Discussing teaching goals

This theme of multivariable problems reveals several levels of requirements for the goals we might assign to our teaching.

Explanations for beginners may be conceived only as a way to make learners familiar with a given phenomenon, and to introduce them to one of the relevant factors implied in this phenomenon. Then, it is suitable to limit oneself to 'soft' arguments. But if learning of how to reason about multidependencies is aimed at, the findings outlined above clearly show that this implies several requirements, in particular that of specifying such a goal. Otherwise, the rather high price to pay will prevent any reasonable teacher from trying. Indeed, there are several factors weighing on the side of an education restricted to monovariable thinking. First, simplifying problems in order to facilitate their analysis is a constituent feature of scientific method. Second, in mathematics education, only monovariable functions are studied, and it is not common to envisage multivariable dependencies. Finally, syllabuses are usually organized along specific contents: *mechanics*, *electric circuits*, or *themes*, *nuclear plants*, *bridges*. Who ever saw 'multivariable problems' in a syllabus?

But there are points that might prompt a teaching community to adopt this goal. Real life is a huge multivariable problem, and if the scientific method consists in isolating good simplified questions, it does not mean that reality spontaneously presents itself in such a form. Hence, we should not let our students go unaware of this. In particular, a highly valuable goal would be to render them capable of deciding that not enough information is available to answer a given question. Moreover, it is not necessary to put into play extremely difficult problems, such as the thermodynamic ones, to introduce such a reflexion. Discussing on whether the area of a rectangle increases or decreases when its width and length are both modified is a good start.

Concluding Remarks

In these two different domains of research about learners' ideas, pedagogical outcomes have been specified according to four items: awareness, vigilance, guidelines for teaching and teaching goals.

A first remark is that this rough classification in fact concerns intricated aspects. This is why the four items have not been dealt with in the same order for each domain: concerning Newton's third law, it is difficult to reach the more subtle level of vigilance without a very clear view about the physics at stake. This clear understanding is more easily got at with the help of the pedagogical tools proposed in the preceding section. Concerning multivariable problems, the 'vigilance' section is placed before the 'pedagogical' one, because the latter is mainly based on the former.

This introduces a second remark. It is worth noting that a domain of research which is transverse with respect to domains of physics, and deals with general ways of reasoning, does not easily lead to pedagogical proposals. The target is diffuse, and the suggestions often seem rather (too?) demanding.

Probably for these reasons, it seems difficult to readily adopt as a pedagogical objective a better mastery of reasoning on multivariable problems. The 'transverse' character of such a goal, as said before, results in the fact that it always follows content-specific goals, which are much better suited to syllabuses, textbooks and teaching plannings. Yet, it might well be that teaching misses the most important points in that way.

References

CHATROUX, Y. (1981). *Physique, classe de Seconde.* (grade 10), 137, Paris: Scodel (Ed.).
CLEMENT, J. (1982). Students' preconceptions in introductory mechanics. *American Journal of Physics*, *50*, 1, pp. 66–71.
Closset, J-L. (1985). Using cognitive conflict to teach electricity. In DUIT, R., JUNG, W. and RHÖNECK, C. (Eds.). *Aspects of understanding electricity.* Kiel: Schmidt and Klaunig, pp. 267–273.
FAUCONNET, S. (1981). *Etude de résolution, de problèmes: Quelques problèmes de même structure en physique.* Unpublished thesis, LDPES, University of Paris 7.
FAUCONNET, S. (1984). Etude de résolution de problèmes analogues. In Delacôte, G. (Ed.). *Research on Physics Education.* La Londes les Maures (CNRS, Paris), pp. 261–269.
FONTAINE, G. and TOMASINO, B. (1990). *Physique, classe de Seconde.* (grade 10), p. 168, Paris, Nathan (Ed.).
LACOURT, J. (1981). *Sciences physiques, classe de troisième.* (grade 9), 96, Paris, Colin (Ed.).
McDERMOTT, L.C. (1984). Research on conceptual understanding in mechanics. *Physics Today*, *37*, pp. 24–32.
MAURINES, L. (1992). Spontaneous reasoning on the propagation of visible mechanical signals. *International Journal of Science Education*, *14*(3), pp. 279–93.
MAURY, J.P. (1989). *La glace et la vapeur, qu'est-ce que c'est?* 27, Paris, Ed. Palais de la découverte.
MENIGAUX, J. (1991). Raisonnements des lycéens et des étudiants en mécanique du solide. *Bulletin de l'Union des Physiciens*, No. 738, pp. 1419–29.
MENIGAUX, J. (1992). Students' reasoning in solid mechanics, submitted to *International Journal of Science Education*, on request from LDPES.
ROZIER, S. (1987). *Le raisonnement linéaire causal en thermodynamique classique élémentaire.* Unpublished thesis, University of Paris 7, on request from LDPES.
ROZIER, S. and VIENNOT, L. (1991). Students' reasoning in thermodynamics. *International Journal of Science Education*, *13*(1), pp. 159–70.
SAISON, A. (1981). *Physique, classe de Seconde.* (grade 10), 95, Paris, Nathan (Ed.).
SHIPSTONE, D.M. (1984). A study of children's understanding of electricity in simple D.C. circuits. *European Journal of Science Education*, *6*(2), pp. 185–198.
VALENTIN, L. (1983). *L'univers mécanique.* Paris: Hermann.
VIENNOT, L. (1979a). Spontaneous reasoning in elementary dynamics. *European Journal of Science Education*, *1*(2), pp. 205–21.
VIENNOT, L. (1979b). *Le raisonnement spontané en dynamique élémentaire.* Paris: Hermann.
VIENNOT, L. (1982). L'action et la reaction sont-elles bien egales et opposées? *Bulletin de l'Union des Physiciens*, No. 640, pp. 479–88.
VIENNOT L. (1985). Analyzing students' reasoning: Tendencies in interpretation. *American Journal of Physics*, *53*, pp. 432–36.
VIENNOT, L. (1989). Bilans de forces et lois des actions réciproques. *Bulletin de l'Union des Physiciens*, No. 716, pp. 951–70.

Chapter 18

Dimensions of Content

Richard T. White

Need for a Theory of Content

It is important to distinguish between types of science content, since different types may best be taught and learned in different ways. We need a theory that depicts the properties of content that matter, and that predicts the teaching and learning procedures that are most effective for each sort.

Attention to a theory of content is overdue. The prominent learning theorists of the 1960s made a start, though they never addressed the differentiation of content into topics. Thus Gagné (1972) distinguished between verbal knowledge, intellectual skills, cognitive strategies, motor skills and attitudes, and further divided the intellectual skills into eight categories, but did not describe how his recommendations for the teaching and learning of each type would be affected by their grouping into coherent topics. Parallel comments could be made about Ausubel's (1968) identification of four forms of meaningful association of propositions, Bruner *et al.'s* (1966) work on concept formation, and Skinner's (1968) on operant conditioning and programmed instruction. Bloom (1968) did consider how bodies of subject matter might be mastered, and emphasized the value of employing a range of teaching procedures, but did not discuss how properties of the content would determine the best method of teaching.

A similar lack of concern for content was evident in research. From the 1950s until well into the 1970s, in studies of the efficacy of teaching procedures such as the popular comparisons of transmissive and discovery methods, content was merely a vehicle. Something had to be learned, any convenient content would do. Even when these studies became more sophisticated, and attribute-treatment interaction studies became common, few if any included content as a variable.

A revolution in appreciation of the role of content in learning was marked by the blossoming in the mid-1970s of research on alternative conceptions of natural phenomena and scientific principles. The first phase of this revolution concentrated on probing students' beliefs. Only much later did researchers begin to try, as yet without clear success, to find how these beliefs might be brought into line with accepted scientific principles. Nor has there been much more than speculation about how alternative conceptions arise in the first place. Consequently, although the research on alternative conceptions has sparked interest in content, it has not yielded clear advice about how to teach different topics.

There is, however, sufficient research to justify a first attempt at deriving principles that relate science content to teaching and to learning, which will constitute the beginnings of a theory of content.

Content is not, of course, the only determinant of how teaching and learning should proceed. There are many other constraints: time, class size, resources, form of assessment, abilities and motivations of the learners and the teacher's relations with them, and the teacher's repertoire of procedures and skill and confidence in using them. Though important, the direct effects of these determinants are beyond the scope of a theory of content. A comprehensive theory would, however, include their interaction with type of content — the learners' prior experiences of the topic and the teacher's knowledge of it must influence how it can best be taught — but such comprehensiveness much wait while initial steps are taken towards identifying properties of content that influence teaching procedures independent of the characteristics of teacher or learners. This chapter is limited to those initial steps.

Properties of Content

Openness to Common Experience

Alternative conceptions in dynamics appear to arise from children's interpretations of common experiences. They see that things do need a force on them to keep moving. Other topics, such as atomic theory, are not experienced at all. Consequently learners are likely to form beliefs about motion, but not about atomic theory, before they have formal lessons on these topics. Different challenges then exist for teacher and learner, and different solutions are likely to be effective.

For topics open to experience, like motion, the learner has to choose between accepting new principles as a veneer to coexist with earlier, competing knowledge, and putting forth effort to resolve the contradictions between the new and the former knowledge. That is, the learner either seals off the new knowledge from his/her extensive experience-based knowledge, or struggles to integrate them. With topics for which there is negligible possibility of direct experience, such as atomic theory, the learner does not enter the classroom with strongly-held beliefs. The learner's problem now is not to resolve conflict but to make sense of the new information. In the case of atomic theory this will happen through acquisition of propositions and images transmitted by the teacher that enable the learner to conceive a macroscopic analogy for the atom. Thus we teach and learn about electronic orbits, in an analogy of the solar system, and energy levels in an analogy of objects at different heights. Since we have no direct experience of atomic structure, we have no competing belief, and it is easy for the teacher to get us to absorb the analogy. The main weakness of our learning may then not be a contradiction in our knowledge, but failure to appreciate that this representation is merely an analogy, with flaws, so that we are confused when it breaks down in instances encountered later. Of course the known half of the analogy is itself a piece of knowledge, and so is formed by experience. Students will have constructed their own meanings for it. If the analogy is not a simple one, then alternative conceptions of it are likely, which will in turn produce diverse understandings of the target topic.

Abstraction

A second important property of content is its level of abstraction, which, though related, is not quite the same dimension as openness to common experience. Quite concrete topics may lie outside experience, or not have much potential for being noticed. An example is the alternation of generations in plants.

Abstraction is not easy to define, since all conceptualizations beyond simple sensations such as pain are artificial. Even concepts that we assume are experienced universally, such as time, are invented. Hence it is better to think of a degree of abstraction for concepts rather than a dichotomy between abstract and concrete. Within the topic motion, the concepts displacement, velocity and acceleration grade in abstraction, as they shade in obviousness to observation. Acceleration is less easy to perceive than change in position, and when it occurs as change in direction at constant speed becomes more abstract. This point about perception of a concept is more subtle than might at first appear. Stavy (chapter 16 in this volume) notes that children are more confident of the concept of liquid than they are of solid. 'Liquidness' is more easy to perceive than 'solidity' because liquids are more similar in appearance than solids are — they all have horizontal surfaces and take the shape of the container, are penetrable and can be poured, all of which are obvious, while solids take all sorts of shapes and forms.

The more abstract the concept, or the less perceptible it is, the less it is open to direct experience, and the less likely that learners will come to the classroom with alternative conceptions for it.

These first two, related properties of openness to experience and abstraction influence the likelihood of students forming beliefs about topics before they encounter formal lessons on them. For open, concrete topics, teachers should find it advantageous to use discussions to elicit students' beliefs, and to design with the class, critical tests of them. Contrast this with a topic for which students have little prior experience (though they might have prior knowledge in the form of facts or procedures). Then the teacher might determine what the students know simply to see where to continue from, but should not expect that the students will already have established conceptions about the content that they are about to encounter. In that case discussion of views in advance might be counterproductive, encouraging students to form a view that may be contrary to the one the teacher plans them to have. Once formed, the contrary view might be difficult to eradicate. When the students have no prior view, the teacher could find it more effective to begin by stating the scientists' view, and to build discussion around that. Transmissive teaching and contrived demonstrations and laboratory exercises are more suitable when there is little common experience related to the topic, though of course must be followed by probes of understanding to check what the students construct from them.

Complexity

Abstraction also needs to be distinguished from complexity. Some topics are relatively coherent, involving only a small number of elements. An example is density, which concerns little more than mass, volume, and the effect of temperature (and for gases, pressure) on it. Other topics include many concepts, and many principles. Sound, for instance, encompasses loudness, pitch, quality, wave motion, frequency, wavelength, intensity, the Doppler effect, reflection, refraction, diffraction, interference, beats, speed in different media, and so forth.

It is probable that different styles of teaching and learning would be required for density and sound. The more complex the content, the greater the need to attend to integrating it and to showing and to perceiving its unity. With sound, for instance, good teaching might involve showing how the molecular model explains all of the phenomena. The teacher might also spend a greater proportion of time on activities such as concept mapping that assist learners to integrate the numerous concepts.

Presence of Alternative Models With Explanatory Power

Whether simple or complex, concrete or abstract, open or closed to common experience, content may differ in whether there are alternative models of good explanatory power. Heat is a topic with such models. The current model of heat as associated with the kinetic energy of molecules displaced a model of it as a fluid, caloric. Erickson (1979 and 1980) found that many high school students hold a caloric view of heat. The caloric model might persist because there is a macroscopic analogy (liquids) in common experience, and because it is effective in explaining many phenomena. The caloric model can invoke an analogy with solution in water to explain why steam is not visible as it comes from a kettle, then appears and disappears.

The existence for heat of a model of explanatory power contrasts with the case of another early theory, that all things are combinations of the four elements fire, air, earth and water. Though there was some fit between this theory and common observation, it strained to explain later phenomena. It does not appear to persist as an alternative conception.

The openness to alternative models of good explanatory power explains why alternative conceptions frequently repeat history. Because science is the development of models of increasingly general explanatory power, many of the discarded ideas were effective in explaining a good range of observations and are useful and satisfying as long as they are kept within that range. The later models are more sophisticated in order to explain a wider range of phenomena, and may be less easy to fit to a common analogy. Heat as a fluid involves a direct link to a concrete experience. Heat as kinetic energy of molecules is explaining heat by a further abstraction that has to be understood in turn.

Presence of Common Words

Topics vary in frequency of use of words that are common in non-scientific language, and in how different are the meanings of those words between technical and common use. This property affects strongly the teaching and learning of the topic.

Bell (1981) has shown how for many people the common use of 'animal' differs from the scientific meaning; 'flower' for most people excludes the biologic flowers of grasses and cauliflowers; 'acceleration' in common use means speeding up, and does not extend to slowing down or changing direction. Some topics, for example dynamics, contain many key terms that are in common use (force, energy, power, momentum, impulse), and those terms differ in how specialized their use is. Force, for instance, has much the same meaning technically and commonly, and so few problems of learning arise; but conservation of energy has different technical and common meanings that have to be clarified. Other topics are dominated by specialized words that are not in common use: for example, genetics, with phenotype and genotype, chromosome, mitosis and meiosis. Here the problem in learning is not only remembering the unusual words but giving them *any* meaning. In other topics there is a mix of specialized and common words, such as light, with diffraction, refraction and spectrum as well as reflection and brightness. The common words in light cause few problems in teaching and learning because they have the same technical and common meaning.

A theory of content might contain the principles that the incidence of alternative conceptions will be greater for topics that employ specialized use of common words, and that rote learning will be more prevalent in topics with a high proportion of unfamiliar words.

Mix of Types of Knowledge

The knowledge people acquire for each of the blocks of content that we name motion, density, redox reactions, transpiration and so forth consists of diverse forms. Elements of knowledge are of different types: propositions, images, episodes, strings, procedures, and motor skills (see White, 1988, for extended descriptions). An important property of content is the emphasis it puts on these different types. Some topics are largely prepositional, for example the properties of the transition elements in the periodic table. Some, such as force and motion, allow for many episodes. Atomic structure and magnetic fields involve a high proportion of images, which often are analogies. Consequently laboratory and field work will be more useful or more important in some topics, while diagrams and students' abilities to understand them will be more crucial in others. The mix of types of knowledge will influence how best the topic may be taught and learned.

Of course, just because a topic *can* be taught in a particular way does not mean it must or will be. A teacher might, for instance, teach density and floating and sinking quantitatively, so that the students acquire intellectual skills of working out the density of an object given its mass and volume and of applying Archimedes' Principle in numerous problems. In contrast is a qualitative emphasis on acquiring physical experiences of hefting, estimating the densities of objects relative to each other, observing things floating, and feeling the upthrust (or apparent lessening of weight) as an object is immersed. In practice the balance between these approaches is a matter of time and resources, and the teacher's judgment of what is appropriate for these students at this time. The point is, though, that topics vary in the mix that is possible.

Teachers may need to ensure that their students can process each type of knowledge, and can link them so that episodes and images illuminate propositions. Probably most teachers and textbook writers have faith in the explanatory power of diagrams, yet it may be that people have as much need to learn how to comprehend diagrams as they do text. Schollum (1983) points out that diagrams in science commonly employ arrows, often to represent vectors, especially forces, but for other meanings as well. Students can be confused by these arrows. The conventional nature of many diagrams needs to be learned. Baird (1984) found students who failed to process graphs — they tried to recall the general shape of the curves without extracting from them the form of the relation being represented. Lowe (1990) found marked differences between experts and novices in meteorology in how they processed weather maps.

The processing of events into a durable memory that illuminates some scientific principle is, like the processing of diagrams, neither automatic nor inevitable. It is, however, important. Mackenzie and White (1982) showed that when students linked memories for events with factual information their recall of the information was greatly enhanced. Training in the processing of experiences and diagrams should benefit learning of topics that contain a mix of types of knowledge.

A theory of content might need to extend to a theory of learners' and teachers' preferences. If learners vary in preference for different types of knowledge, as White and Gunstone (1980) found, then each is likely to learn some topics better than others. Similarly, teachers may differ in their capabilities for handling different types of knowledge — some may work best with facts, others diagrams, others events — so that some topics they teach well, others poorly.

With most topics some latitude exists for the mode of presentation. Organic chemistry, for instance, can be taught with words, diagrams and models. Teachers can shift the emphasis on each of these in accord with their skills and their students' preferences.

Demonstrable Versus Arbitrary

Even though two topics might involve similar proportions of the various types of knowledge, there may be differences within each type that affect how the topics should be taught. This includes how alternative conceptions should be addressed. Propositions, for instance, may be demonstrable or arbitrary. The action of dilute acids on metals is demonstrable, as are the propositions that the moon revolves round the earth and that green plants are phototropic. But the statement that electric current flows from the positive terminal of a cell is arbitrary. Statements about classification systems are arbitrary. By definition, arbitrary propositions rest on authority and cannot be proved by demonstration. This does not mean that they are irrational, only that an element of choice is present. Thus the European division of the year into four seasons is defensible, even though alternative systems of three, or other number, of seasons could be followed. Arbitrary propositions vary in rationality or in usefulness, but probably most of those in science have some reason behind them. Hence a teacher might explain the advantages of the (arbitrary) division between metals and non-metals, or could take care to explain to students how 'animal' can have different scientific and social meanings.

Social Acceptance

Another property of propositions is their degree of social acceptance, which is coming to be an issue in curriculum design and teaching. Many topics in science are contentious: creation and evolution, population and industrial development, power generation and consumption of fossil fuels, population control. Propositions within these topics are open to debate. In some cases the debate may be ended by evidence. This should happen soon, for instance, for the proposition that 'planets are common around stars', less soon for 'life exists on other planets', and perhaps never for 'extra-sensory perception exists'.

We need principles of how to deal with the learning, and unlearning, of propositions that vary in social acceptance as well as in arbitrariness-demonstrability.

Extent of Links

Quite often in discussing the papers in this volume, participants referred to how extensively a topic could be linked to other content. Thus when Symons, Brass and Odgers (chapter 13 in this volume) found that students had trouble with connecting human heart and lungs in their models of circulation of the blood, this could be related to single circulation systems of reptiles and to hole-in-the heart babies, and eventually to fluid transfer in plants.

Topics vary in how self-contained they are versus how extensively they are related to other topics. Energy is a key concept in many topics in physics, chemistry, biology and earth science, as, to a lesser extent, is electricity. Friction, quantum mechanics and genetic inheritance are more limited. Pervasive topics would need to be taught differently from restricted ones.

Teaching must not miss opportunities for encouraging students to perceive links between topics. Apart from the deeper understanding that will follow, connectedness is at the heart of science. Carr *et al.* (chapter 11 in this volume) argue that connectedness is one of the criteria for the value of explanations of the natural world. Without connectedness, science is not a system capable of further advance, but a collection of eclectic trivia.

Emotive Power

One property of content that requires more thought than I have so far been able to give it is its emotive power. Different topics are likely to arouse different types, as well as different intensities, of emotion. Of course the emotions that they actually arouse will be influenced strongly by the personalities and experiences of the learners and the skill of the teacher, but some topics will have more potential than others for producing wonder, delight or excitement, or antipathy, disgust or boredom. I imagine there is more potential fun in floating and sinking than in rusting, in the properties of hydrogen than in the gas laws — but I may be mistaken about that.

Implications of differences in emotive type and power for teaching and learning remain to be discovered. Should teachers capitalize on emotion, or would that make it harder to encourage learning of the less arousing topics? Are alternative conceptions more common, or harder to shift, in emotive topics? Such questions have rarely if ever been discussed.

Development of the Theory

I have described several properties of content that may affect how it should be taught and learned, and in some cases how best to tackle alternative conceptions. Although most of the foregoing theory consists of descriptive propositions, there are some predictions, for example 'discussion of students' beliefs will be advantageous for topics that are open to experience and concrete, and harmful for topics that are closed to experience and abstract'. Further development of the theory should involve (a) lists of the predictions; (b) deduction of more predictions from further consideration of the described properties; (c) specification of further properties; and (d) merging of the theory of content with theories of learning and of instruction.

References

AUSUBEL, D.P. (1968). *Educational psychology: A cognitive view.* New York: Holt, Rinehart and Winston.

BAIRD, J.R. (1984). *Improving learning through enhanced metacognition.* Ph.D. thesis, Monash University.

BELL, B.F. (1981). When is an animal, not an animal? *Journal of Biological Education, 15*, pp. 213–8.

BLOOM, B.S. (1968). Learning for mastery. *Evaluation Comment*, Los Angeles: Center for the Study of Evaluation of Instructional Programs, UCLA.

BRUNER, J.S., OLVER, R.R. and GREENFIELD, P.M. *et al.* (1966). *Studies in cognitive growth.* New York: Wiley.

ERICKSON, G.L. (1979). Children's conceptions of heat and temperature. *Science Education, 63,* pp. 221–30.

ERICKSON, G.L. (1980). Children's viewpoints of heat: A second look. *Science Education, 64*, pp. 323–36.

GAGNÉ, R.M. (1972). Domains of learning. *Interchange, 3*, pp. 1–8.

LOWE, R.K. (1990). Diagram information and its organisation in memory: Exploring the role of skill and experience. *Research in Science Education, 20,* pp. 191–9.

MACKENZIE, A.A. and WHITE, R.T. (1982). Fieldwork in geography and long term memory structures. *American Educational Research Journal, 19*, pp. 623–32.

SCHOLLUM, B. (1983). Arrows in science diagrams: Help or hindrance for pupils? *Research in Science Education, 13*, pp. 45–9.

SKINNER, B.F. (1968). *The technology of teaching.* New York: Appleton Century Crofts.

WHITE, R.T. (1988). *Learning Science.* Oxford: Blackwell.

WHITE, R.T. and GUNSTONE, R.F. (1980). Converting memory protocols to scores on several dimensions. Australian Association for Research in Education. *Annual Conference papers,* pp. 486–93.

Postscript

Peter J. Fensham

Much has been learnt in the last fifteen years about the learning of science and hence about how science can be taught to improve the quality of that learning. This book has presented many of these exciting research findings and given examples of these good classroom practices in relation to a variety of science content.

We are, however, not so naive to imagine that this is any more than necessary knowledge for the improvement of science education. Science teachers and teacher educators will construct their own meanings about what we have written. The process of producing the book was, as we indicated in the Preface, a constructive experience for all of us as authors. The meanings others will draw and the judgments they will make about using them will be coloured very much by the educational contexts they are in. Each of our working and living contexts shape and constrain what we regard as practicable and worth making the effort to try to implement.

The physical contexts in which many science teachers and teacher educators work often impose quite severe constraints. On the other hand, I have often observed quite different responses in teachers who share essentially the same physical constraints. Some see only teacher-centred, transmissive teaching of science as possible, while some create minds-on, active learning by their students.

The curriculum climate in which teachers carry out their science teaching is a major determining factor that is, in many ways, more pervasively influential (usually constraining) than are physical limitations. If the knowledge worth learning is defined by an externally-set examination consisting of multiple choice items, then it is particularly difficult for teachers to encourage their students to achieve deeper understanding of concepts and phenomena. If there are no rewards for extended investigations by students of open-ended scientific problems, it is hard for teachers to allow students the time these sorts of learnings of the nature of science require. If the sociotechnical context of everyday society is merely seen as motivational and not content to be learnt about, then why should science teachers do other than use examples of it to ease the boredom of their students? These aspects of the curriculum climate of science education are often beyond the individual teacher to influence. Collectively, however, through their science teacher associations they can exert (and have done from time to time) political pressures that lead to changes in the climate. Science teacher educators and curriculum developers both as individuals and collectively do have opportunities to contribute to changes in the climate. Elsewhere (Fensham, 1991), I have argued that science teacher educators are now potentially a very influential group in determining how the knowledge of worth in science education is determined.

The authors of this book and others who value constructivist ideas in science education have, I believe, some further contributions to make to the creation of more helpful curriculum climates. There is now an urgent need for the attention of these researchers to turn to reformulations of two other influential components of science education besides teachers. These are science text books and the tests that are used to assess learning in science education.

Text books can be and are sources for learning science whether or not a student's individual teacher is able or not so able. Current texts have not usually been written with constructivist principles in mind. Indeed, it is not at all clear what a text that would encourage constructive learning would be like. Given the pervasiveness of text books in most countries, developing such texts is a challenging research project to undertake.

Tests, or examinations, are as we all know very powerful indicators of what is the knowledge worth learning. Assessment of science learning, particularly comparative assessments are politically growing in significance rather than declining. At present, most of the assessment instruments that exist, from the Science Olympiads to the multiple choice alternative sets used by so many teachers, essentially measure, and hence affirm, the recall of closed, rather low level science information and algorithmic skills. Knowing more of this rather superficial learning is the criterion used to differentiate students, rather than measures which explore the relative depth of their learnings. There is a desperate need to develop, and establish confidence in new sorts of assessment instruments in science that themselves will act as rewards and encouragements for more constructive science education. White and Gunstone (1992) have made a start by collecting imaginative procedures from research projects and describing how they can be applied in classrooms.

It is to these two huge tasks that I hope some of our energies will turn in the next few years.

References

Fensham, P.J. (1991). Science and technology. In Jackson, P.W. (Ed.). *Handbook of Research on Curriculum*. New York: Macmillan, pp. 789–829.

White, R.T. and Gunstone, R.F. (1992). *Probing Understanding*. London: Falmer Press.

Notes on Contributors

Hilary Asoko taught biology and general science in schools in England and overseas for eleven years before her appointment as Lecturer in Science Education at the University of Leeds. She is involved mainly with in-service and pre-service training programmes for primary teachers and is a member of The Children's Learning in Science Research Group. Her particular interest is in the development of teaching approaches to promote conceptual understanding in science.

Kate Brass came to Monash in 1989 after teaching high school science and working on curriculum development projects in Ireland and Zimbabwe. At Monash she has worked on a variety of science education research projects concerned with the quality of science learning and teaching. She enjoys rambling in the Australian bush and photographing Australian flora and fauna.

Malcolm Carr is Director of the Centre for Science and Mathematics Education Research at the University of Waikato, Hamilton, New Zealand. He taught chemistry at the tertiary level for twenty-eight years, and has combined this interest with science education for the last twelve years. The theatre, music and walking on beaches complement and enrich his academic activities.
Co-authors.: Miles Barker, Beverley Bell, Fred Biddulph, Alister Jones, Valda Kirkwood and John Pearson have worked with the Learning in Science Projects at the University of Waikato, and contributed greatly to the debates and sharing of ideas leading to this paper. David Symington made the same large contribution from Victoria College in Melbourne, Australia.

Rosalind Driver is Professor of Science Education at the University of Leeds. Her longstanding interest in children's ideas about natural phenomena was stimulated during her experience as a physics teacher in Africa and South America. Since her appointment at Leeds in 1973, she has contributed to the training of science teachers and the work of the Assessment of Performance in Science team (1979–1982). In 1983 she initiated the Children's Learning in Science Research Group which she continues to direct. Her professional interests are in the learning of science, teacher development and the public understanding of science.

Reinders Duit has been a member of the physics education group of the Institute for Science Education at the University of Kiel, Germany, since 1969 and a professor for physics eduction since 1990. His main research interests have been students' difficulties in learning basic science concepts. He is also working at setting constructivist ideas of science teaching into practice, for instance, via a new physics textbook for secondary schools. Listening to music, singing in a choir, canoeing and riding a bicycle help to avoid the onesidedness of being a constructivist science educator only.

Maureen Duke has been a primary school teacher for twenty years. As well as her interest in science, she has researched children's attitudes to mathematics and has explored a number of strategies for implementing an integrated curriculum for multi-age groups. Outside of teaching she is actively involved in local community work and pursues interests in craft, photography and painting.

Jonathan Emberton is Head of the Biology department at Ryburn Valley High School which is located 30 miles west of Leeds. He first became aware of the work of the CLIS Research Group through an extended in-service training course in 1986 and has since been involved in a number of research projects. He has particular interests in teaching and learning about ecology and, outside his professional life, finds time to compete nationally in orienteering events.

Gaalen Erickson is a professor in the Department of Mathematics and Science Education at the University of British Columbia where he has worked with pre-service teachers and graduate students since 1975. Prior to that he taught science at the high school level in Alberta and England. His professional concerns include the analysis of how school pupils learn more about their world and how beginning science teachers learn how to teach. Too infrequently he makes forays into the mountains for skiing, cycling, and hiking.

Peter Fensham was the first Professor of Science Education in Australia after teaching university chemistry for ten years. From 1967–1992 he held that position at Monash University but was also Dean of the Faculty of Education from 1982–1988. He has been involved in Australia in science education, in curriculum policy, and in social justice in schools and higher education. Tai Chi, bush walking and the social Gospel are important sustaining influences.

Dick Gunstone is an Associate Professor in the Education Faculty at Monash University. Before joining Monash in 1974 he taught physics and maths in Victorian high schools. Currently his teaching and research embrace science education, metacognition and professional development of pre- and in-service teachers. The Skeptics, historical aspects of science, sporting trivia, photography and Ashley Brilliantianism are sources of inspiration.

Peter Häussler worked as a physicist at Mainz University, Germany and at the University of California in Berkeley before joining the Institute for Science Education (IPN) in Kiel, Germany, in 1970. Since 1980 he has also been a member of the science faculty of Kiel University, being appointed as a professor for physics education in 1988. He has been involved in various empirical studies on problem solving, aims for physics education, long-term effects of physics instruction and interests towards physics. Playing the violin, painting and canoeing are his favorite pastimes.

David Hawkins is Distinguished Professor of Philosophy, Emeritus, University of Colorado. He was first director of the Elementary Science Study (ESS) 1960–64, and a MacArthur Fellow 1981–86. His work has been primarily in Science, Philosophy of Education and in practical efforts to improve the teaching of science for young children. Home life, hiking, reflections on number theory, and carpentry provide relaxation and enjoyment.

Wendy Jobling has taught in Victorian government primary schools for a number of years. She is interested in how children learn and are taught Science and Technology Studies, and has been a member of the Monash Children's Science Group for several years. Out of school hours she is a Technology Studies In-service Course facilitator, and is also studying for the degree of Doctor of Education at Monash University. When time allows, Wendy enjoys Wing Chun, skiing and sailing on Victoria's beautiful Gippsland Lakes.

Cliff Malcolm is currently co-ordinator of the Australian Science Curriculum and Teaching Program and a leader in the development of National Curriculum. He is an Associate of the Education Faculty at Monash University. In the Victorian Ministry of Education from 1983–1990 he led the state development of science education policy, curriculum frameworks and student materials and was active in teacher support. Before joining the Ministry of Education, he taught Physics, Primary Science and Environmental Science at Melbourne State College. His hobbies centre on family activities and also tennis, scuba diving and guitar.

Sylvie Rozier is a physics teacher for students preparing for French 'Grandes Ecoles'. She graduated in 1988 with a thesis about students' reasoning in thermodynamics. Currently, her research is on causal reasoning in chemistry. Two young children contribute to a well-occupied life.

Telsa Rudd is a primary school teacher with several years of experience. She has always enjoyed working and learning alongside her young students, watching them develop their interests. Teaching Science, Technology and Maths has always interested Telsa as a means of helping primary students develop their learning skills, enabling them to become independent learners and thinker. When Telsa has time to relax, she enjoys riding her horse, Biggles, practising 'Chester Burtwhistle' (her cello) and going for walks and sailing with her husband, Mike.

Phil Scott is Co-ordinator of the Children's Learning in Science Research Group at the University of Leeds. He first became involved in the work of the group whilst teaching science in a Leeds city high school and is particularly interested in investigating how research into children's learning might be drawn upon to inform teaching practice. Family, music, Sunderland soccer and fell running figure highly in his non-professional life.

Ruth Stavy is an Associate Professor in Science Education at the School of Education at Tel Aviv University. Before joining Tel Aviv University in 1974 she did research as a biochemist at the Weizman Institute of Science. From 1974–1985 she was engaged in elementary school science curriculum development at the Israel Science Teaching Center at Tel Aviv University. She has been a member of the academic staff at the School of Education since 1985, acting as the head of the Life Science Education Program. Her main interest is in the development of scientific concepts in students. Some of her other interests are the history of science, reading literature and travelling.

Anne Symons is Head of the Science Department at Korowa Anglican Girls' School in Melbourne, Victoria. She teaches Biology and General Science from years 7 to 12. She is also involved in working with primary teachers at her school as they pilot a new national primary science program developed by the Australian Academy of Science. Anne has been a member of the Monash Children's Science Group for a number of years and has been interested in developing teaching strategies that maximize student learning.

Laurence Viennot, after five years of research in astrophysics, moved to didactics of physics in 1971. Now a professor at the University of Paris, France, she teaches 'pure' physics and didactics of physics. Her research is more specifically focused on students' and teachers' reasoning in physics. She is responsible for the 'Laboratoire de Didactique de la Physique dans l'Enseignement Superieur' in Paris, and is a member of the French National Curriculum Council. Although really bad at modern dance, she has not yet given up practising it.

Jim Wandersee is the Associate Professor of Biology Education in the Graduate Faculty at Louisiana State University. Before affiliating with Louisiana State University, he taught high school sciences for ten years and college biology for eleven years. His primary research interest is the graphic representation of scientific knowledge and its effects on learning. Desert backpacking, the culture of *Cyperus alternifolius*, the music of Aaron Copland, the work of Leonardo Da Vinci, and professional cartooning are welcome distractions from professorial work.

Mike Watts has taught secondary science in comprehensive schools in London and Kingston, Jamaica. He gained his doctorate at the University of Surrey before becoming Regional Project Leader for the Secondary Science Curriculum Review, a national science project in the UK. He is currently Reader in Education at Roehampton Institute in London, with research interests in children's conceptual development, open-ended problem solving and teachers' professional development. He is also hopelessly devoted to the guitar and clarinet music, copious amounts of beer and the Welsh rugby team.

Richard White taught general science, physics and chemistry for ten years in high schools before joining Monash University in 1971. He has been Professor of Educational Psychology since 1981. His professional interest is in the quality of learning, which has led him into research on learning hierarchies, episodic memory, cognitive structure and metacognition. Active participation in sailing and painting and more passive pursuits of reading, listening to music, computer games and admiring Australian plants supplement his professional life.

Merlin Wittrock is a professor, chairman of the Faculty, and head of the division of educational psychology of the Graduate School of Education at UCLA, where he has been a faculty member since 1960. His research interests centre on the psychological processes involved in the teaching of reading comprehension, and in the learning of science and mathematics. He has published widely on learning and teaching, including editing the *Handbook of Research on Teaching. Third Edition*.

Acknowledgements

The writing workshop and this subsequent volume have involved many people. We particularly wish to acknowledge the expertise of

— Daphne Attwood, who handled all the organization of the workshop
— Sandra Bosmans, who produced from a variety of sources the final typed manuscript
— Sarah Harrington and Claude Sironi, who produced a number of the diagrams in the book
— Heidi Kass, Leo MacDonald and Ian MacDonald constructed the index

Author Index

Subject Index

9 780750 702218

For Product Safety Concerns and Information please contact our EU
representative GPSR@taylorandfrancis.com
Taylor & Francis Verlag GmbH, Kaufingerstraße 24, 80331 München, Germany

www.ingramcontent.com/pod-product-compliance
Ingram Content Group UK Ltd.
Pitfield, Milton Keynes, MK11 3LW, UK
UKHW022220190926
475702UK00003B/454
* 9 7 8 1 1 3 8 1 6 5 8 0 9 *